LIMING ACIDIC SURFACE WATERS

Harvey Olem

Library of Congress Cataloging-in-Publication Data

Olem, Harvey.
Liming acidic surface waters / author, Harvey Olem.
p. cm.
Includes bibliographical references and index.
ISBN 0-87371-243-9
1. Acid pollution of rivers, lakes, etc. 2. Water quality management. 3. Lime. 4. Neutralization (Chemistry) I. Title
TD427.A27043 1991
628.1'68--dc20 90-13458
CIP

LEWIS PUBLISHERS, INC.
121 South Main Street, Chelsea, Michigan 48118

PRINTED IN THE UNITED STATES OF AMERICA

Harvey Olem is President of Olem Associates, an environmental consulting firm currently producing several publications that provide decisionmakers with scientific assessments and research results to aid in improved protection of the aquatic environment.

The author earned a Ph.D. in Civil Engineering and M.S. in Environmental Pollution Control from Pennsylvania State University and received a B.S. from Tufts University in Medford, Massachusetts. He has over 17 years experience in environmental protection activities and has authored over 75 articles and reports on environmental topics.

Prior to establishing Olem Associates he was a projects manager and environmental engineer with the Tennessee Valley Authority, a researcher for the Institute for Research on Land and Water Resources, and an instructor for the Pennsylvania Department of Community Affairs. He first became interested in acid rain in 1981 while he was at Tennessee Valley Authority. He was appointed to the Interagency Task Force on Acid Precipitation as the TVA representative to the aquatic effects working group and eventually developed a keen interest in aquatic liming.

He is a registered professional engineer in Massachusetts, Pennsylvania, and Virginia and is an officer, director, or committee chair for a number of professional organizations such as the Water Pollution Control Federation, North American Lake Management Society, and the Alliance for Environmental Education. He lives in Herndon, Virginia with his wife Sheila and son Michael.

Acknowledgments

Many persons provided valuable assistance during the preparation of this book. Several researchers in the United States, Canada, Sweden, Norway, and the United Kingdom sent me many of the articles and reports cited here. Their assistance was greatly appreciated.

This information was released by the National Acid Precipitation Assessment Program (NAPAP) as number 15 in the State of Science and Technology series. I thank NAPAP for encouraging the outside publication of this report.

I thank Timothy B. Adams, Joseph Bergin, Gillian M. Booth, Michael Bowman, David J.A. Brown, Robert W. Brocksen, Paul A. Bukaveckas, Charles T. Driscoll, Paul H. Eschmeyer, James Fraser, Patricia Irving, Penelope Kellar, Richard H. Lepley, Michael Marcus, Lynn Moore, Donald Porcella, Carole Shriner, Howard A. Simonin, Perry Suk, Harald U. Sverdrup, Walter Warnick, Chantale Wong, and Amy Woodis. I thank the anonymous peer reviewers whose constructive and helpful suggestions improved the manuscript.

I especially thank Eva Thornelof, Swedish Environmental Protection Board, for translating sections of a number of Swedish reports into English. I also thank Steinar Sandoy, Norway Directorate for Nature Management, for Norwegian translations. These translations allowed the incorporation of important information that has not before appeared in the English literature and would not otherwise have been included in this report.

Finally, I thank R. Kent Schreiber of the U.S. Fish and Wildlife Service and his staff—James M. Brown, William A. Hartman, and Rita F. Villella—for their timely and constructive comments during the review process. Dr. Schreiber arranged for financial support, assisted in determining the scope and contents of the report, and, most important, believed in my ability to produce a comprehensive state of the science and technology report on liming.

Foreword

Water chemistry in aquatic environments has been significantly altered by both natural and anthropogenic disturbances. Glaciation—a natural phenomenon—filled many watersheds with a mantle of glacial till that affected lakes and streams, making them infertile and acidic. Human activities such as mining and use of fossil fuels have exacerbated this natural acidity, increasing the number of impaired aquatic systems.

To increase productivity and restore their usefulness for agriculture, aquaculture, industry, and habitats for wildlife, these waterbodies are treated with a variety of chemicals and other materials. The most commonly used naturally occurring material is limestone, which is applied in a process called "liming."

Fishery managers have long recognized the value of increasing waterbody productivity through intensive management. Since liming is one of the few effective techniques available to deal with acidity problems in lakes and streams, "Liming Acidic Surface Waters" by Dr. Harvey Olem will be a significant addition to the fishery manager's repertoire of methods to increase productivity. These resource managers know that sport fishing is important to the public and that liming makes both scientific and economic sense.

Liming of waterbodies is a controversial process, however. Critics fear that such an efficient, cost-effective method of increasing surface water pH will lull the public and lessen demand to control the sources of acidification. In any case, source control will be gradual and may only restore about half of currently acidic surface waters. It is clear that aquatic liming should be used as an independent, additional control.

This is another reason that Dr. Olem's book is welcome. A comprehensive treatment of the complexities of chemistry and biology, current information on the types, sizes, and locations of these waterbodies, and the most complete guide to date on the problems and solutions associated with liming, it will facilitate dealing with acidified surface waters.

As a fishery scientist and as president of the Sport Fishing Institute, I applaud Dr. Olem's timely and authoritative book. "Liming Acidic Surface Waters" will ultimately benefit not only the growing number of sport fishermen—60 million in the United States alone—but also the nonfishing public, who will enjoy wider recreational opportunities.

Gilbert C. Radonski
President
Sport Fishing Institute

Preface

This document describes the science and technology for aquatic liming—a method for improving the water quality of acidic surface waters to restore or enhance fisheries. The report is a comprehensive compilation of years of research in North America and Europe by dozens of scientists. Several mitigation technologies—including those that have only been proposed—are critically evaluated, along with the effects of liming on water chemistry and aquatic biota. Through these evaluations, the state of the science and technology of aquatic liming is identified for the reader.

Whole-lake liming is now recognized as a valuable management tool for acidic surface waters and their fisheries. However, some liming technologies are considered experimental and will need further evaluation. Distinctions between technologies are included—as is the distinction between liming acidic surface waters and reducing acidifying emissions.

The manuscript has undergone two rounds of peer review by four anonymous experts selected by the National Acid Precipitation Assessment Program (NAPAP) and a critical review by Dr. Harald Sverdrup of the Lund Institute of Technology, Lund, Sweden. It was also examined by other experts selected by the author and by members of the NAPAP Interagency Science and Interagency Policy Committees from the Environmental Protection Agency, National Oceanic and Atmospheric Administration, Department of Agriculture, Department of Energy, Department of the Interior, Council on Environmental Quality, and the Tennessee Valley Authority.

The manuscript was revised on the basis of these reviews and of comments received at the NAPAP International Conference held February 11–16, 1990. The reviewers' recommendations resulted in several positive changes.

Abbreviations

ALaRM .... Acid Lake Reacidification Model
ANC .... Acid neutralizing capacity
EPA .... U.S. Environmental Protection Agency
EPRI .... Electric Power Research Institute
LLI .... Living Lakes, Inc.
NAPAP .... National Acid Precipitation Assessment Program
SLiM .... Soil Liming Model

Common and Scientific Names of Fishes Mentioned in this Book

Scientific Name	Common Name
Alosa aestivalis	Blueback herring
Alosa pseudoharengus	Alewives
Coregonus albula	Cisco
Esox lucius	Northern pike
Etheostoma exile	Iowa darter
Etheostoma nigrum	Johnny darter
Micropterus dolomieui	Smallmouth bass
Morone americana	White Perch
Morone saxatilis	Striped bass
Notropis cornutus	Common shiner
Notropis simus	Bluntnose minnow
Oncorhynchus mykiss, formerly *Salmo gairdneri*	Rainbow trout
Perca flavescens	Yellow perch
Perca fluviatillis	European perch
Pimephales promelas	Fathead minnow
Rhinichthys atratulus	Blacknose dace
Rutilus rutilus	Roach
Salmo salar	Atlantic salmon
Salmo trutta	Brown trout
Salvelinus alpinus; *Salvelinus salvelinus*	Arctic char
Salvelinus fontinalis	Brook trout
Salvelinus namaycush	Lake trout
Semotilus atromaculatus	Creek chub
Stizostedion vitreum vitreum	Walleye
Umbra limi	Central mudminnow

Contents

List of Figures

List of Tables

Definitions

Absence - when used in the context of the presence or absence of a fish species or other organisms, species not caught during sampling (generally using a standardized sampling regime for all waters sampled) are defined as absent.

Abundance - the number of organisms per unit area or volume.

Accuracy - the difference between the approximate solution obtained by using a numerical model and the exact solution of the governing equations (or a known standard concentration), divided by the exact solution (or known standard concentration); the closeness of the measured value to the true value of an analyte.

Acid anion - negatively charged ion that does not react with hydrogen ion in the pH range of most natural waters.

Acid-base chemistry - the reaction of acids (proton donors) with bases (proton acceptors). In the NAPAP context, this means the reactions of natural and anthropogenic acids and bases, the result of which is described in terms of the **pH** and **acid neutralizing capacity** of the system.

Acid cation - hydrogen ion or metal ion that can hydrolyze water to produce hydrogen ions, e.g., ionic forms of aluminum, manganese, and iron.

Acid mine drainage - runoff with high concentration of metals and sulfate and high levels of acidity resulting from the oxidation of sulfide minerals that have been exposed to air and water by mining activities.

Acid neutralizing capacity (ANC) - the equivalent capacity of a solution to neutralize strong acids. The components of ANC include weak bases (carbonate species, dissociated organic acids, alumino-hydroxides, borates, and silicates) and strong bases (primarily, OH^-). In the National Surface Water Survey, as well as in most other recent studies of acid-base chemistry of surface waters, ANC was measured by the Gran titration procedure.

Acidic deposition - transfer of acids and acidifying compounds from the atmosphere to terrestrial and aquatic environments via rain, snow, sleet, hail, cloud droplets, particles, and gas exchange.

Acidic episode - an **episode** in a water body in which **acidification** of surface water to an **acid neutralizing capacity** less than or equal to 0 occurs.

Acidic lake or stream - a lake or stream in which the **acid neutralizing capacity** is less than or equal to 0.

Acidification - the decrease of **acid neutralizing capacity** in water or **base saturation** in soil caused by natural or anthropogenic processes.

Acidified - pertaining to a body of water that has experienced a decrease in **acid neutralizing capacity** or a soil that has experienced a reduction **in base saturation**.

Acidophilic - describing organisms that thrive in an acidic environment.

Adaptation - a change in the sensitivity of an organism to an environmental stress or factor over time.

Alkalinity - for this report, the equivalent sum of $HCO_3^- + CO^{2-} + OH^-$ minus H^+, i.e., buffering conferred by the bicarbonate system; the terms *ANC* and *alkalinity* are sometimes used interchangeably. ANC includes alkalinity plus additional buffering from dissociated organic acids and other compounds. (See **acid neutralizing capacity**).

Analyte - a chemical species that is measured in a water sample.

Analytical model - mathematical model in which the solution of the governing equations is obtained by mathematical analysis, as opposed to numerical manipulation.

Anion - a negatively charged ion.

Anion-cation balance - a method of assessing whether all ions have been accounted for and measured accurately; in an electrically neutral solution, such as water, the total charge of positive ions (cations) equals the total charge of negative ions (**anions**).

Anion deficit - the concentration of measured **cations** minus measured **anions**; usually the result of unmeasured organic anions or analytical uncertainty and often used as a surrogate for organic anion concentration.

Anion exchange and adsorption - a reversible process occurring in soil by which anions are adsorbed and released.

Anion reduction - the process by which NO_3^- is reduced to N_2O or N_2 (nitrate reduction or denitrification) and SO_4^{2-} is reduced to S^{2-} (**sulfate reduction**). Nitrate reduction and sulfate reduction can be mediated by plants during growth; the reduced N and S are assimilated into the growing plant. Denitrification and dissimilatory sulfate reduction are mediated by bacteria in anoxic zones in soils, sediment, or the water columns.

Anthropogenic - of, relating to, derived from, or caused by humans or related to human activities or actions.

Base cation - an alkali or alkaline earth metal cation (Ca^{2+}, Mg^{2+}, K^+, Na^+).

Base cation buffering - the capacity of a watershed soil or a sediment to supply base cations (Ca^{2+}, Mg^{2+}, K^+, Na^+) to receiving surface waters in exchange for acid cations (H^+, Al^{3+}); may occur through cation exchange in soils or weathering of soil or bedrock minerals.

Base cation supply - (1) the pool of **base cations** (Ca^{2+}, Mg^{2+}, K^+, Na^+) in a soil available for exchange with acid cations. The base cation pool is determined by the cation exchange capacity of the soil and the percentage of exchange sites occupied by base cations. (2) The rate at which base cations can be supplied to buffer incoming acid cations; this rate is determined by the relative rate of mineral weathering, the availability of base cations on exchange sites, and the rate of mobile anion leaching.

Base saturation - the proportion of total soil **cation exchange capacity** that is occupied by exchangeable **base cations**, i.e., by Ca^{2+}, Mg^{2+}, K^+, and Na^+.

Battery-powered doser - small stream liming device typically power by a 12-V battery.

Bedrock - solid rock exposed at the surface of the earth or overlain by saprolites or unconsolidated material.

Behavioral avoidance - the act of avoiding or withdrawing from an environmental stimulus.

Benthic invertebrates -organisms lacking a spinal column that live in association with the lake or stream bottom. Discussions in this book are limited to benthic **macroinvertebrates**, i.e., organisms large enough to be retained by a 0.595-mm sieve.

Benthic - bottom zones or bottom-dwelling organisms in water bodies.

Bioassay - measurement of the response of an organism or group of organisms upon exposure to in situ environmental conditions or simulated environmental conditions in the laboratory; also referred to as **toxicity test**.

Biological community - a general collective term to describe the varieties of organisms (multiple species) living together in a given ecosystem.

Biological effects - changes in biological (organismal, populational, community-level) structure or function in response to some causal agent; also referred to as **biological response**.

Biological population - an assemblage of organisms of the same species inhabiting a given ecosystem.

Biological recovery - an observed (or projected) sustained improvement in indicators of biological condition at the species, population, or community level, in response to actual (or simulated) decreased acidic conditions.

Biological response variables - quantifiable measurements of biological response (see **variable, biological effects**).

Biological significance - the quality of being important in maintaining the structure or function of biological populations or communities.

Biomass - the total quantity of organic matter in units of weight or mass.

Buffer intensity - the change in **pH** per unit addition of **strong acids** or **strong bases**.

Calibration - process of checking, adjusting, or standardizing operating characteristics of instruments and model appurtenances on a physical model or coefficients in a mathematical model with empirical data of known quality. The process of evaluating the scale readings of an instrument with a known standard in terms of the physical quantity to be measured.

Catchment - see **watershed**.

Cation - a positively charged ion.

Cation exchange - the interchange between a cation in solution and another cation on the surface of any surface-active material such as clay or organic matter.

Cation exchange capacity - the sum total of exchangeable cations that a soil can adsorb.

Cation leaching - movement of cations out of soil, in conjunction with mobile anions in soil solution.

Cation retention - the physical, biological, and geochemical process by which cations in watersheds are held, retained, or prevented from reaching receiving surface waters.

Chelator - a class of compounds, organic and inorganic, that can bind metal ions and change their acid-base chemistry and biological availability.

Chemical precipitation - the chemical reaction of aqueous (dissolved) solutes to form a solid phase of precipitate.

Chronic acidification - see **long-term acidification**.

Circumneutral - close to neutrality with respect to **pH** (neutral pH = 7); in natural waters, pH 6–8.

Comparability - a measure of data quality that assesses the similarity within and among data sets.

Conceptual model - simplified or symbolic representation of prototype or system behavior and responses.

Conductance (or conductivity) - the ability of a substance (e.g., aqueous solution) to carry an electrical current. For natural waters, conductance is closely related to the total concentration of dissolved ions. Conductance is usually measured under very specific conditions (see **specific conductance**).

Confidence limits - a statistical expression, based on a specified probability, that estimates the upper or lower value (limit) or the interval expected to contain the true population mean.

Contracid™ - commercially available base treatment technique involving sediment treatment with sodium carbonate.

Convergence - state of tending to a unique solution. A given scheme is convergent if an increasingly finer computational grid leads to a more accurate approximation of the unique solution. Note that a numerical method may sometimes converge on a wrong solution.

Decomposition - the microbially mediated reaction that converts solid or dissolved organic matter into its constituents (also called *decay* or *mineralization*).

Denitrification - biologically mediated conversion of nitrate to gaseous forms of nitrogen (N_2, NO, N_2O); denitrification occurs during decomposition of organic matter.

Detritus - dead and decaying organic matter originating from plants and animals.

Discharge areas - geographic portions of a watershed where there is surface flow, particularly during periods of stormflow or snowmelt.

Dispersant - chemical agent added to limestone to improve dissolution and minimize settling.

Dissolution efficiency - ratio between the dissolved amount of neutralizing material in water and the total added amount over a given unit of time.

Dissolved inorganic carbon - the sum of dissolved carbonic acid, carbon dioxide, bicarbonate, and carbonate in a water sample.

Dissolved organic carbon - organic carbon that is dissolved or unfilterable in a water sample (0.45-μm pore size in the National Surface Water Survey).

Diversion well - a cylindrical concrete structure containing crushed limestone through which stream water is diverted to allow mechanical grinding of limestone and subsequent neutralization.

Dose-response relationship - the association between the environmental concentration to which an organism (or population) is exposed and the expected response or biological effect.

Dose - the quantity of material added per unit volume or per unit of time.

Doser - any mechanical device designed to continuously treat acidic flowing waters by the addition of base materials.

Drainage basin - see **watershed**.

Drainage lake - a lake that has a permanent surface water inlet and outlet.

Dry deposition - transfer of substances from the atmosphere to terrestrial and aquatic environments by the gravitational settling of large particles and turbulent transfer of trace gases and small particles.

Dry-powder doser - an automated device that stores dry limestone powder and dispenses either dry powder or slurried powder.

Dynamic model - a mathematical model in which time is included as an independent variable.

Dystrophic lake - a lake low in nutrients yet highly colored with dissolved humic organic matter.

Empirical model - representation of a real system by a mathematical description based on experimental or observational data.

Episodes - a subset of hydrological phenomena known as *events*. Episodes, driven by rainfall or snowmelt, occur when **acidification** takes place during a **hydrologic event**. Changes in other chemical variables, such as aluminum and calcium, are frequently associated with episodes.

Episodic acidification - the short-term decrease of **acid neutralizing capacity** from a lake or stream. This process has a time scale of hours to weeks and is usually associated with hydrological events.

Equivalent - unit of ionic concentration, a mole of charge; the quantity of a substance that either gains or loses 1 mol of protons or electrons.

Eutrophication - a process of accelerated aquatic primary production in response to nutrient enrichment, which ultimately can result in oxygen depletion and changes in biological community structure and function.

Evapotranspiration - the process by which water is returned to the air through direct evaporation or transpiration by vegetation.

Fingerlings - post-larval juvenile fish; "a small fish no longer than a finger."

Fishery yield - the quantity of fish harvested or available for harvest.

Forecast - to estimate the probability of some future event or condition as a result of rational study and analysis of available data.

Gran analysis - a mathematical procedure used to determine the equivalence points of a titration curve for **acid neutralizing capacity**.

Groundwater - water in a saturated zone within soil or rock.

Groundwater flow-through lake - a **seepage lake** that receives a substantial amount of groundwater. Although there is no clear distinction between this type of lake and a groundwater recharge lake, groundwater flow-through lakes have been operationally defined as having silica concentrations ≥15 μeq/L.

Groundwater recharge lake - a **seepage lake** that receives little or no groundwater input, but discharges water to the groundwater system. Operationally, groundwater recharge lakes have been defined as having silica concentrations <1.0 mg/L or potassium concentrations <15 μeq/L.

Headwater lake - uppermost lake in a river system.

Hydraulic residence time - a measure of the average amount of time water is retained in a lake basin. It can be defined on the basis of inflow/lake volume, represented as RT, or on the basis of outflow (outflow/lake volume) and represented as T_w. The two definitions yield similar values for fast-flushing lakes, but diverge substantially for slow-flushing seepage lakes.

Hydrologic(al) event - pertaining to increased water flow or discharge resulting from rainfall or snowmelt.

Hydrologic(al) flow paths - surface and subsurface routes by which water travels from where it is deposited by precipitation to where it drains from a **watershed**.

Hydrology - the science that treats the waters of the earth — their occurrence, circulation, and distribution; their chemical and physical properties; and their reaction with their environment, including their relationship to living things.

Initial conditions - given values of dependent variables or relations between dependent and independent variables at the initiation of a computation.

Inorganic aluminum - the sum of free aluminum ions (Al^{3+}) and dissolved aluminum bound to inorganic ligands; operationally defined by **labile monomeric aluminum**.

Interactive model - numerical model that allows interaction by the modeler during application.

Ionic strength (I) - a measure of the interionic effect resulting primarily from electrical attractions and repulsions between various ions. $I = 1/2\Sigma C_i Z_i^2$ for all ions i (cations and anions), where C_i is the concentration of ion i and Z_i is the valence.

Jar test - a laboratory test for determining effectiveness of various neutralizing agents conducted by adding candidate agents to known quantities of the water to be treated. After 24 hr of stirring, the final **pH** of the sample is plotted against dosage.

Labile monomeric aluminum - operationally defined as aluminum that can be retained on a cation exchange column and measured by one of the two extraction procedures used to measure **monomeric aluminum**. Labile monomeric aluminum is assumed to represent inorganic monomeric aluminum (Al_i).

Lentic - referring to standing waters, as in ponds and lakes.

Limestone barrier - barrier of limestone aggregate placed across a stream channel to neutralize acid water.

Liming - the addition of any base materials to neutralize surface water or sediment or to increase **acid neutralizing capacity**.

Littoral zone - the shallow, near-shore region of a body of water; often defined as the band from the shoreline to the outer edge of the occurrence of rooted vegetation.

Long-term acidification - the decrease of **acid neutralizing capacity** in a lake or stream on a seasonal or longer time frame.

Lotic - referring to flowing waters, as in streams.

Macroinvertebrates - invertebrate organisms, usually benthic, large enough to be retained by a 0.595-mm sieve.

Macrophytes - macroscopic forms of aquatic vegetation.

Mathematical model - model using mathematical relationships to represent the **prototype**.

Mean pH - calculated from the mean hydrogen ion concentration of individual samples.

Mechanistic model - **mathematical model** in which actual ecosystem processes are explicitly included (see **deterministic model**).

Mineral acids - inorganic acids, e.g., H_2SO_4, HNO_3, HCl, H_2CO_3 (see **strong acids** and **weak acids**).

Mineral weathering - dissolution of rocks and minerals by chemical and physical processes.

Mitigation - generally described as amelioration of adverse impacts caused by acidic deposition at the source (e.g., emissions reductions) or the receptor (e.g., lake liming). For the purposes of this report, amelioration of acidic conditions in surface waters to preserve or restore fisheries and the supporting aquatic community.

Mixing zone - for the purposes of this report, an area of stream between the dosing device and the beginning of the target area, where the neutralizing agent mixes.

Mobile anions - anions that flow in solutions through watershed soils, wetlands, streams, or lakes without being adsorbed or retained through physical, biological, or geochemical processes.

Model - an abstraction or representation of a **prototype** or system, generally on a smaller scale.

Monomeric aluminum - aluminum that occurs as a free ion (Al^{3+}), simple inorganic complexes (e.g., $Al(OH)_n^{3-n}$, AlF_n^{3-n}), or simple organic complexes, but not in polymeric forms; operationally, extractable aluminum measured by the pyrocatechol violet method or the methyl-isobutyl ketone method (also referred to as the *oxine* method) is assumed to represent **total monomeric aluminum**. Monomeric aluminum can be divided into labile and nonlabile components by using cation exchange columns.

Natural acids - acids produced within terrestrial or aquatic systems through natural, biological, and geochemical processes, i.e., not a result of acidic deposition or deposition of acid precursors.

Nitrification - oxidation of ammonium to nitrite or nitrate by microorganisms. A byproduct of this reaction is H^+.

Nitrogen fixation - biological conversion of elemental nitrogen (N_2) to organic N.

Nonlabile monomeric aluminum - operationally defined as aluminum that passes through a cation exchange column and is measured by one of the two extraction procedures used to measure monomeric aluminum; assumed to represent organic monomeric aluminum (Al_o).

Numerical model - mathematical model in which the governing equations are not solved analytically. The governing equations are solved approximately; discrete numerical values and arithmetic operations are used to represent the variables involved.

Nutrient cycling - the movement or transfer of chemicals required for biological maintenance or growth among components of the ecosystem by physical, chemical, or biological processes.

Oligotrophic lake - describes a lake with little organic matter and little biological activity.

Organic acids - acids possessing a carboxyl (-COOH) group or phenolic (C-OH) group; includes fulvic and humic acids.

Organic aluminum - aluminum bound to organic matter, operationally defined as that fraction of aluminum determined by colorimetry after the sample is passed through a strong cation exchange column.

Pelagic zone - referring to open-water areas not directly influenced by the shore or bottom.

Perched seepage lakes - see **groundwater recharge lake**.

Periphyton - as used here, plants that live closely attached to underwater surfaces (e.g., bottom sediments or macrophytes).

pH - the negative logarithm of the hydrogen ion activity. The **pH** scale extends from 1 (most acidic) to 14 (most alkaline); a difference of one **pH** unit indicates a 10-fold change in hydrogen ion activity.

Physiography - the study of the genesis and evolution of land forms; a description of the elevation, slope, and aspect of a study area.

Plankton - small (mostly microscopic) plant or animal species that spend part or all of their lives carried passively by water currents.

Pool - (1) in streams, a relatively deep area with low velocity; (2) in ecological systems, the supply of an element or compound, such as exchangeable or weatherable cations or adsorbed sulfate, in a defined component of the ecosystem.

Population - for this report, (1) the total number of lakes or streams within a given geographical region or the total number of lakes or streams with a given set of defined chemical, physical, or biological characteristics; or (2) an assemblage of organisms of the same species inhabiting a given ecosystem.

Population-level response - a change in the characteristics of a biological population (e.g., a change in population abundance) in response to some causal agent.

Precision - a measure of the capacity of a method to provide reproducible measurements of a particular **analyte** (often represented by variance).

Predict - to estimate some current or future condition within specified confidence limits on the basis of analytical procedures and historical or current observations.

Prediction - a calculation or estimation of some current or future condition on the basis of historical or current observation, experience, or scientific reason (see **predict**).

Primary production - the conversion of inorganic nutrients and carbon into organic matter by plants and bacteria that use sunlight or, more rarely, inorganic compounds by oxidation.

Profundal - the **benthic** portion of a lake below the zone of light penetration (e.g., deeper than the **littoral** zone).

Project - to estimate future possibilities based on rational study and current conditions or trends.

Prototype - the full-sized structure, system, process, or phenomenon being modeled.

Quality assurance - a system of activities for which the purpose is to provide assurance that a product (e.g., data base) meets a defined standard of quality with a stated level of confidence.

Quality control - steps taken during sample collection and analysis to ensure that data quality meets the minimum standards established in a **quality assurance** plan.

Reacidification - see **acidification**.

Receptor - as used here, the location of mitigation treatments; can be a lake, stream, or watershed.

Recovery - see **biological recovery**.

Reduction/oxidation (redox) reactions - reactions in which substances gain or lose electrons, i.e., in which substances are converted from an oxidized to a reduced oxidation state and vice versa.

Refugia - areas or zones in aquatic ecosystems in which environmental conditions are relatively nonstressful for biological organisms.

Representativeness - a measure of data quality; the degree to which sample data accurately and precisely reflect the characteristics of a population.

Reproductive potential - the expected number of next-generation female eggs produced by a current-generation female egg. In a **population** at steady state, the reproductive potential is 1. The lower bound is zero for a **population** that is completely unable to reproduce.

Respiration - the biological oxidation of organic carbon with concomitant reduction of external oxidant and the production of energy. In aerobic respiration, O_2 is reduced to CO_2. Anaerobic respiration processes use NO_3^- (**denitrification**), SO_4 (**sulfate reduction**), or CO_2 (**methanogenesis**).

Retention time - the estimated mean time (usually expressed in years) that water remains in a lake before it leaves the system (see **hydraulic residence time**).

Riffle - in streams, a relatively shallow area with high-velocity water flow extending across a stream, in which water flow is broken at the surface.

Rotary drum - water-powered liming device for streams in which limestone aggregate is ground in a revolving drum.

Scenario - one possible deposition sequence after the use of a control or mitigation strategy and the later effects of this deposition sequence.

Secchi disk depth - a measure of the transparency of water by the use of a round disk that is lowered into a water body until it can no longer be seen.

Sediment injection - mitigation technique in which sodium carbonate is injected into lake bottom sediment (see also **Contracid**™).

Seepage lake - a lake with no permanent surface water inlets or outlets. Seepage lakes are sometimes divided into two categories: **groundwater recharge lakes** and **groundwater flow through lakes**.

Sensitive - tending to experience **capacity effects** or **intensity effects**, as a result of changes in acidic deposition; or biological effects as a result of a change in acid-base chemistry.

Sensitivity - the tendency of a system to experience **capacity effects** or **intensity effects** as a result of changes in acidic deposition, or biological effects as a result of a change in **acid-base chemistry**; a measure of the degree to which a **model's** output responds to small changes in an input value.

Sensitivity analysis - test of a **model** in which the value of a single **variable** or **parameter** is changed, and the impact of this change on the dependent variable is observed.

Short-term acidification - see **episode**.

Silo - a limestone storage bin.

Simulation - description of a **prototype** or system response to different conditions or inputs with a **model** rather than by actually observing the response to the conditions or inputs.

Simulation model - **mathematical model** that is used with actual or synthetic input data, or both, to produce long-term time series or predictions.

Slurry - a liquid mixture containing suspended limestone particles.

Species richness - the number of species occurring in a given aquatic ecosystem, generally estimated by the number of species caught when a standard sampling technique is used.

Specific conductance - the conductivity between two plates with an area of 1 cm^2 across a distance of 1 cm at 25°C.

Steady state - the condition that occurs when a property (e.g., mass, volume, concentration) of a system does not change with time. This condition requires that sources and sinks of the property be in balance (e.g., inputs equal outputs; production equals consumption).

Steady-state model - a model in which the variables under investigation are assumed to reach equilibrium and are independent of time.

Stream order - a method of categorizing streams based on their position in the drainage network. First-order streams are permanent streams with no permanent tributaries. Higher order streams are formed by the confluence of two or more streams of the next lower stream order.

Stream reach - a **stream segment** of relatively uniform morphology. The term is scale dependent in that it may refer equally to segments of stream that have uniform macrohabitat features (e.g., a "riffle reach") or a relatively uniform mix of macrohabitat characteristics (e.g., a uniform distribution of riffles, pools, plunges); in the National Stream Survey – Phase I, segments of the stream network are represented as blue lines on 1:250,000-scale USGS maps. Each reach is defined as the length of stream between two blue-line confluences.

Stream segment - usually used to refer to the length of stream channel between two stream tributary confluences. Often used to refer to a length of stream with uniform physical and morphological characteristics. In these two usages, **stream segment** is scale dependent and is very similar in meaning to **stream reach**.

Strong acids - acids with a high tendency to donate protons or to completely dissociate in natural waters, e.g., H_2SO_4, HNO_3, HCl^-, and some **organic acids** (see **acid anions**).

Strong bases - bases with a high tendency to accept protons or to completely dissociate in natural waters, e.g., NaOH.

Subpopulation - any defined subset of the target **population**.

Subtle effects - changes that are difficult to detect because of their small magnitude (relative to natural background variability) or because they occur slowly over a long period of time.

Sulfate reduction - (1) the conversion of sulfate to sulfide during the decomposition of organic matter under anaerobic conditions (dissimilatory sulfate reduction) and (2) the formation of organic compounds containing reduced sulfur compounds (assimilatory sulfate reduction).

Surface area-weighted mean particle diameter - average diameter weighted to an individual particle's surface area.

Surficial geology - characteristics of the Earth's surface, especially consisting of unconsolidated residual, colluvial, alluvial, or glacial deposits lying on the bedrock.

Tonne - metric ton or 1000 kg; convert to tons by multiplying by 0.9072.

Total monomeric aluminum - the simple unpolymerized form of aluminum present in inorganic or organic complexes.

Toxicity test - see **bioassay**.

Turnover - the interval time in which the density stratification of a lake is disrupted by seasonal temperature variation, resulting in the entire water mass becoming mixed.

μeq/L - 10^{-6} equivalents per liter; 1 g equivalent weight of an acid is the quantity of acid that can donate 1 mol of protons (H^+) to a base.

Unsaturated flow - flow of water through the voids in rock or soil at a pressure less than atmospheric, i.e., flow in response to gravitational and capillary forces through the unsaturated zone between the land surface and the water table.

Validation - comparison of **model** results with a set of **prototype** data not used for **verification**. Comparison includes the following: (1) using a data set very similar to the verification data to determine the validity of the model under conditions for which it was designed; (2) using a data set quite different from the verification data to determine the validity of the model under conditions for which it was not designed but could possibly be used; and (3) using postconstruction prototype data to determine the validity of the **predictions** based on model results.

Variable - a quantity that may assume any one of a set of values during analysis.

Verification - check of the behavior of an adjusted model against a set of **prototype** conditions.

Volume-weighted mean particle diameter - average particle size based on the mass of limestone fraction and assuming that all of the particles are spherically shaped.

Water-powered doser - mechanical device that uses the energy of running water to operate a mechanism to dispense limestone powder.

Watershed - the geographic area from which surface water drains into a particular lake or point along a stream.

Watershed liming - neutralization of surface waters by applying limestone to the terrestrial portion of the watershed. The areas may include wetlands, recharge areas (dry soils), and discharge areas.

Weak acids - acids with a low proton-donating tendency that tend to dissociate only partly in natural waters, e.g., $H_2CO_3^-$, $H_4SiO_4^-$, and most **organic acids** (see **acid anions**).

Weak bases - bases with a low proton-accepting tendency that tend to dissociate only partly in natural waters, e.g., HCO_3, $Al(OH)_4$.

Wet deposition - transfer of substances from the atmosphere to terrestrial and aquatic environments by precipitation, e.g., rain, snow, sleet, hail, and cloud droplets. Droplet deposition is sometimes referred to as occult deposition.

Wetland - area that is inundated or saturated by surface or groundwater at a frequency and duration sufficient to support a prevalence of vegetation typically adapted for life in saturated soil.

Wet-slurry doser - an automated device that stores preslurried limestone into streams.

Yearling (fish) - fish 1 year old (age 1) in its second year of life.

Young-of-the-year (fish) - fish less than 1 year old (age 0).

1

Introduction

"Snow and adolescence are the only problems that disappear if you ignore them long enough."

Earl Wilson

THE NEED FOR MITIGATION

Mitigation of acidic conditions in surface waters (primarily by liming) is desirable because acidic waters are detrimental to aquatic organisms. High concentrations of the ions H^+ and Al^{n+} in acidic waters adversely affect ion regulation (a condition known as osmoregulatory failure) in aquatic organisms. The principal detrimental effect on fish and other aquatic organisms is the leaching of sodium chloride from body fluids (Baker et al., 1990a). Elevated H^+ and Al^{n+} concentrations may also cause gas-transfer problems and lead to asphyxiation. Acidic surface waters also tend to be nutrient-deficient waters that, even in the absence of acid-related stress, may not support significant fisheries.

In general, higher order biological groups are less able to adapt to increasing acidity (Baker et al., 1990a). The ranking of biological groups in order of decreasing sensitivity to acidity for community-level effects is fish > benthic invertebrates and zooplankton > phytoplankton > microorganisms.

The general types of biological changes expected to occur with increasing surface water acidity at 0.5 pH intervals is summarized in Table 1.1. Increasing acidity is shown in 0.5 pH units and values refer to levels in summer or during fall overturn. The effects cited assume that water below pH 5.2 would contain elevated aluminum. Calcium concentrations are assumed to be fairly low (50–250 μeq/L). A more complete discussion of the assumptions used to prepare this information was given by Baker et al. (1990a). The conditions shown in the table are likely to be seen in surface waters that are candidates for mitigation.

Table 1.1 General Summary of Biological Changes Anticipated after Surface Water Acidification, Expressed as a Change in pH (Source: Baker et al., 1990a)

pH Decrease	General Biological Effects
6.5 to 6.0	Small decrease in species richness of phytoplankton, zooplankton, and benthic invertebrate communities resulting from the loss of a few highly acid-sensitive species, but no measurable change in total community abundance or production
	Some adverse effects (decreased reproductive success) may occur for highly acid-sensitive fish species (e.g., fathead minnow, striped bass)
6.0 to 5.5	Loss of sensitive species of minnows and dace, such as blacknose dace and fathead minnow; in some waters decreased reproductive success of lake trout and walleye, which are important sport fish species in some areas
	Visual accumulations of filamentous green algae in the littoral zone of many lakes, and in some streams
	Distinct decrease in the species richness and change in species composition of the phytoplankton, zooplankton, and benthic invertebrate communities, although little if any change in total community biomass or production
	Loss of a number of common invertebrate species from the zooplankton and benthic communities, including zooplankton species such as *Diaptomus sicilis, Mysis relicta, Epischura lacustris*; many species of snails, clams, mayflies, and amphipods, and some crayfish
5.5 to 5.0	Loss of several important sport fish species, including lake trout, walleye, rainbow trout, and smallmouth bass; as well as additional nongame species such as creek chub
	Further increase in the extent and abundance of filamentous green algae in lake littoral areas and streams
	Continued shift in the species composition and decline in species richness of the phytoplankton, periphyton, zooplankton, and benthic invertebrate communities; decreases in the total abundance and biomass of benthic invertebrates and zooplankton in some waters
	Loss of several additional invertebrate species common in oligotrophic waters, including *Daphnia galeata mendotae, Diaphanosoma leuchtenbergianum, Asplanchna priodonta*; all snails, most species of clams, and many species of mayflies, stoneflies, and other benthic invertebrates
	Inhibition of nitrification
5.0 to 4.5	Loss of most fish species, including most important sport fish species such as brook trout and Atlantic salmon; few fish species able to survive and reproduce below pH 4.5 (e.g.,

Table 1.1 General Summary of Biological Changes Anticipated after Surface Water Acidification, Expressed as a Change in pH (Source: Baker et al., 1990a) (continued)

pH Decrease	General Biological Effects
	central mudminnow; yellow perch; and, in some waters, largemouth bass)
	Measurable decline in the whole-system rates of decomposition of some forms of organic matter, potentially resulting in decreased rates of nutrient cycling
	Substantial decrease in the number of species of zooplankton and benthic invertebrates and further decline in the species richness of the phytoplankton and periphyton communities; measurable decrease in the total community biomass of zooplankton and benthic invertebrates in most waters
	Loss of zooplankton species such as *Tropocyclops prasinus mexicanus*, *Leptodora kindtii*, and *Conochilis unicornis*; and benthic invertebrate species, including all clams and many insects and crustaceans
	Reproductive failure of some acid-sensitive species of amphibians such as spotted salamanders, Jefferson salamanders, and the leopard frog

MITIGATION: SOURCE AND RECEPTOR

Two methods are available to alleviate acidic conditions in surface waters that have been caused by acidic deposition. One strategy, restricting fossil fuel combustion emissions of NO_x and SO_2, deals with the problem at the source (Office of Technology Assessment, 1984; Herrick, 1987). Once emission reductions are called for, however, it may take a decade or longer for the reduction to be effected. It is likely that such reductions may not result in aquatic conditions suitable for fish restoration for a considerable time beyond the achievement of emission reductions (Thornton et al., 1990). Also, essentially continuous efforts are needed in the future to suppress emissions.

The other strategy involves the receptor and results in mitigation of acidic conditions in surface waters by the addition of alkaline materials or by other methods within the lake, stream, or watershed. This strategy, unlike mitigation at the source, produces only temporary effects if lakes and streams continue to be subject to acidic deposition, but may be necessary if fish restoration is a short-term objective. The implementation of mitigative techniques targeted at receptors and their results are usually immediate. Mitigation at the receptor may no longer be needed for some surface waters once emissions levels are lowered. Other surface waters may require continued mitigation because of acidic inputs from soils or other manmade or natural sources.

In this book, the techniques, costs, benefits, and overall effectiveness of mitigation strategies at the receptor (i.e., lake or stream) are described and no comparisons between mitigation strategies at the source (e.g., emission controls) and the receptor are made. It is assumed that mitigation of acidic surface waters will continue to be independent of emissions reductions; thus the techniques, effects, and effectiveness need to be reviewed and assessed.

OBJECTIVES

The objectives of this work are to

- summarize the various mitigation strategies that have been used or investigated to alleviate acidic conditions in surface waters,
- describe the physical, chemical, and biological effects of these strategies, and
- bring together and synthesize the technology (mitigation strategies) and the science (effects of mitigation).

These objectives are met through the use of all available literature known to the author on mitigation of acidic surface waters, including many outside the United States.

Chapter 1 provides a summary of the history of aquatic liming practices throughout the world. The discussion is divided between aquatic liming practices begun before about 1970 and after 1970. Results of most of the liming projects conducted after 1970 are discussed in more detail in applicable sections. This division was arbitrary, but effectively distinguishes between early liming practices that were not necessarily conducted to alleviate the effects of acidic deposition and major liming projects during and after 1970 that, for the most part, were conducted in response to acidification that was believed to be due to acidic deposition.

Chapter 2 describes the various strategies that have been used for mitigation of acidic surface waters. The most widely used mitigation strategy, and the one covered in greatest detail in this

work, is the addition of base material (liming). Other mitigation strategies covered here are nutrient addition, stocking of acid-tolerant fish strains, and pumping of alkaline water (e.g., groundwater) into surface waters. Included in the discussion of mitigation strategies based on liming are the neutralizing agents available; the techniques used in lake, stream, and watershed applications; dosage calculations; reacidification models; and the cost of neutralizing materials and application methods.

The physical, chemical, and biological effects of liming practices are described in Chapter 3. The focus of effects research is first summarized for recent liming experiments conducted as part of major mitigation programs in the United States. The next discussion provides information, by variable, on the findings of studies reporting the physical and chemical effects of liming. The final discussion in Chapter 3 includes information on the effects of liming on biological processes (nutrient cycling, decomposition, and primary productivity) and biota (plankton, macrophytes, benthic macroinvertebrates, and fish).

Chapter 4 describes the resource management aspects of mitigation, including a discussion of where mitigation of acidic surface waters should be considered, criteria and procedures for selection of candidate systems, and estimates of the number of systems in the United States that are candidates for mitigation. Estimates are provided for Canada, Sweden, and Norway for comparison.

Because numerous surface waters are mentioned in this work, listings of all lakes and flowing waters specifically discussed in the text are included in Appendix A. The years in which the lake or stream was treated are provided, along with characteristics of the water body (surface area or average flow) and information on the treatments (material, dosage, and application method).

Appendix B applies selected modeling approaches for estimating the required limestone dose to treat diverse situations in lakes used as examples. The optimum doses selected by each approach are then compared. A comparison of methods for predicting reacidification rates is also included.

Appendix C is a detailed summary of the results of major mitigation projects in the United States. The results of some major experimental liming projects outside the United States are included in Appendix D. Appendices C and D supplement information contained in Chapter 3 on the effects of liming.

BACKGROUND

As mentioned previously, the aquatic liming practices presented here are arbitrarily divided between aquatic liming practices that occurred before and after 1970.

Aquatic Liming Practices Initiated Before 1970

Reducing the acidity of surface waters has been practiced since the early 1900s, primarily to enhance fish production in dystrophic systems or to neutralize acidity formed by pyrite oxidation as the result of coal mining. Table 1.2 summarizes the major liming activities begun before 1970.

The first reported attempt to lime waters that were believed to be acidified by acidic deposition was conducted in Norway in the 1920s (Rosseland and Skogheim, 1984). Mortalities of salmon eggs and fry were high in many hatcheries in southern Norway. Initial attempts at neutralization were conducted by using a variety of materials, including sodium hydroxide, sodium carbonate, and calcium chloride. Limestone (calcium carbonate) was found to be most successful in hatcheries; where widely used, limestone filters raised the pH from 5.2 to 6.0, increasing survival. By the mid-1930s, limestone filters were introduced in 80% of the hatcheries in Norway.

Liming of farm ponds has been conducted for many years, primarily to improve the productivity of the ponds. Liming to increase the productivity and to enhance the hatching of fish was reported as early as 1909 in Sweden (Swedish Fishing Journal, 1909). The addition of lime to pond sediments was reported to be particularly popular for fish ponds in the southeastern United States (Boyd, 1974; Boyd and Scarsbrook, 1974; Arce and Boyd, 1975; Boyd, 1976; Boyd, 1982a,b). Lime introduced in this way increases the availability of phosphorus for plant growth, which in turn increases fish production. Pond sediments are typically analyzed to determine how much lime is required to neutralize their acidity (Boyd and Cuenco, 1980). These practices remain popular today in farm pond management.

The chemical neutralization of mineral acidity in streams and lakes affected by acid mine drainage has been practiced in Appalachia and other coal regions of the world for many years (Pearson and McDonnell, 1978; Olem and Longaker, 1981; Olem,

Table 1.2 Major Liming Activities Begun Before 1970

Program	Location	Focus of Program or Project	Beginning Date	Reference
Norwegian hatcheries	Norway	Improve or restore hatcheries for production of salmon eggs and fry	1920s	Rosseland and Skogheim, 1984
Southeastern U.S. ponds	Alabama Georgia	Enhance fish production, decrease acidity of sediments	1950s	Boyd, 1974
Acid mine drainage	Appalachia	Neutralize lakes and streams affected by acid mine drainage	1960s	Pearson and McDonnell, 1975a,b
New York Department of Environmental Conservation	New York (Adirondacks)	Improve or restore fisheries in the Adirondack region	Late 1950s	Blake, 1981; Simonin et al., 1988
Massachusetts Department of Fisheries and Wildlife	Massachusetts	Improve or restore fisheries in three small ponds	Late 1950s	Bergin, 1984; General Research Corp., 1984

Table 1.2 Major Liming Activities Begun Before 1970 (continued)

Program	Location	Focus of Program or Project	Beginning Date	Reference
West Virginia Department of Natural Resources	West Virginia	Improve or restore fisheries in three lakes and two streams	1960s	Zurbuch, 1963, 1984
Dystrophic ponds (upper midwest region of the United States)	Wisconsin, Michigan (Upper Peninsula)	Enhance fish production	1940s	Hasler, 1951

1982; Knauer, 1986). Early reports of treatments used to neutralize acid mine drainage date back to the early 1900s in Appalachia. Chemical neutralization became more widespread during the 1960s with the advent of more stringent legislation and the enforcement of laws designed to reduce or halt environmental damage caused by mining (Lovell, 1973; Penn Environmental Consultants, 1983).

Although acid mine drainage and acidic deposition can both contribute to the acidification of surface waters, the results of mitigation strategies are not strictly comparable because of the inherent differences in the water quality and volume of surface waters (Fraser and Britt, 1982). Surface waters affected by acid mine drainage are typically much lower in pH and higher in ionic strength and toxic metal concentrations than are waters believed to be neutralized in lakes and streams affected by acidic deposition. Also, the volume of water to be neutralized in lakes and streams affected by acidic deposition is often larger than that requiring treatment as a result of acid mine drainage.

Many of the techniques used to treat acid mine drainage have been adapted or slightly modified for use in mitigating the effects of acidic deposition. The major difference in techniques is the lack of acceptance by the mining industry of the use of limestone, although its use in the treatment of acid mine drainage has been recommended in many situations (Wilmoth, 1977). Calcium hydroxide (hydrated lime) or calcium oxide (calcined lime) have been the materials of choice in acid mine drainage treatment. Most early treatments for nonmining situations in New York consisted of the use of hydrated lime, but the practice had been almost entirely discontinued by 1976 because its use increases the possibility of pH shock to aquatic biota (Simonin, 1988; Simonin et al., 1988).

A limited number of aquatic liming projects were conducted in the United States from the 1940s to the 1960s on dilute acidic waters not affected by mine drainage (Fraser and Britt, 1982). These were primarily state programs that operated on limited budgets. Next are summarized such activities in five states: New York, Massachusetts, West Virginia, Wisconsin, and Michigan (Fraser and Britt, 1982; Fares and Kinsman, 1984). Some analytical techniques used in these early studies (e.g., determination of pH and alkalinity) may be considered unacceptable today.

New York

The New York State Department of Environmental Conservation had an extensive early liming program, treating 27 surface waters before 1970 (Simonin et al., 1988). Between 1959 and 1963, 25 ponds in the Adirondack region of New York were limed. Of these, 19 ranging from 0.2 to 90 ha in surface area, were in Franklin County in the northern Adirondacks; 3 were clearwater acidic ponds; and the other 16 were darkwater bog ponds. The ponds were treated by applying calcium hydroxide directly to the water surface by boat. Pretreatment pH ranged from 3.6 to 4.8, and the dosage was calculated to yield a posttreatment pH of 6.8. Treated ponds were successfully stocked with brook trout and rainbow trout. Many of the ponds were allowed to reacidify over a period of several years before reliming, thereby stressing the remaining fish (Gloss et al., 1988a,b).

An operational liming program in New York began in 1964 and continues today with the inclusion of a small experimental component (Kretzer and Colquhoun, 1984; Simonin, 1988; Simonin et al., 1988). From 1964 to 1969, six waters were treated, of which four were reapplications to lakes treated previously. For treatments before 1970 hydrated lime was primarily used, but agricultural limestone was used in later treatments. Treatments to accessible areas were conducted by boat or barge, and remote sites were treated from a helicopter. Additional information on the operational program is presented in a subsequent section, and results are included in Appendix C.

Massachusetts

The Massachusetts Department of Fisheries and Wildlife began limited liming of ponds in 1957 to improve or restore fish productivity (Bergin, 1984). Four ponds with a total surface area of 23.8 ha were treated before 1970. Berry Pond, a 1.2-ha fish culture pond in North Andover, was treated twice in 1957 with 1.1 tonnes of agricultural limestone. Other treatments include a 2.4-ha pond at Ft. Devens in 1963 and two ponds in Plymouth—one in 1964 and another in 1969. The program continued on a limited scale during the 1970s and 1980s.

West Virginia

The West Virginia Department of Natural Resources is carrying out a limited liming program with an emphasis on streams (Zurbuch, 1963, 1984; Fares and Kinsman, 1984). Two impoundments (total surface area, 28 ha) have been treated. Spruce Knob Lake was limed in 1958 and has been treated periodically thereafter; Summit Lake was also limed in 1965 and was retreated annually thereafter.

West Virginia in the late 1950s actively began pursuing effective techniques for the neutralization of acidic streams. Zurbuch (1984) described early attempts involving the placement of bags of hydrated lime in small tributaries with the hope that the material would slowly dissolve and neutralize acidity in the main stream. Because this approach was ineffective due to insufficient contact between the lime and the water, further testing of other methods was attempted.

A mechanical doser system was tested in the late 1950s, but the method was not considered adequate because of (1) the need for daily maintenance and (2) the erratic pH levels that resulted. The next system examined was the placement of limestone in the streambeds. Reactivity of the stone decreased with time and little neutralization occurred. In the early 1960s, a rotary drum filled with limestone aggregate was tested on Condon Run, a tributary of Otter Creek in the Monongahela National Forest near Elkins. The initial success of this method led to other installations, as well as the development of a self-feeding system in the 1970s. The application and effectiveness of this type of system are described later.

Wisconsin

In Wisconsin, liming is reported to have begun in the 1940s to enhance fish production in dystrophic ponds and lakes (Hasler et al., 1951; Johnson and Hasler, 1954; Kitchell and Kitchell; 1980; Garrison, 1984). Garrison (1984) summarized the results of lime additions to five Wisconsin lakes between 1950 and 1969. The first treatments involved additions of hydrated lime to Cather Lake and Turk Lake in northwestern Wisconsin, two brownwater lakes with reduced transparency. In addition to hydrated lime additions to the lake surfaces, agricultural limestone was spread near the shores of both lakes. The treatments were successful in raising the pH to about 7 and increasing the transparency of the lakes. Two additional brownwater lakes in north central Wisconsin were treated in the late 1950s, again with hydrated lime; alkalinity and transparency were increased.

Gabriel Lake was treated in 1969 with hydrated lime, agricultural limestone, superphosphate, and ammonium nitrate to increase the fertility of this small, clearwater lake. After treatment the pH increased from 5.8 to 7.2. This lake was monitored periodically until early 1980, when the pH was still about 7 and the alkalinity was almost five times greater than it was before liming (140 μeq/L vs. 30 μeq/L).

Michigan

Lake liming experiments were first reported in the Upper Peninsula of Michigan in the 1940s (Waters, 1956; Waters and Ball, 1957). These early investigators raised the pH in 1954 in two bog lakes from pH 5 to more than pH 9 by applying hydrated lime. Algae in the lake were adversely affected by the high pH during the first year after treatment. Much of the lime became immobilized in the hypolimnion and thus was unavailable for improving the productivity of the lakes. A third lake, a brownwater lake with peat in the sediment, was treated first in 1943 with hydrated lime and later with various dosages of pebble limestone, additional hydrated lime, and crushed dolomite in an attempt to increase productivity. Although the lake was initially circumneutral, alkalinity increased from 120 μeq/L to 320 μeq/L. The investigators found that growth of young-of-the-year yellow perch increased significantly.

Stross and Hasler (1960) reported the results of liming Peter-Paul Lake that began in 1951 and continued until 1954. This hourglass-shaped lake consisted of two basins connected by a narow, shallow canal. The lake was separated in 1951 by an earthen dam; one side (Peter Lake) received the treatments and the other side (Paul Lake) was used as a natural control. Calcium hydroxide was added by boat, resulting in increases in pH (from about 4.7 to 7.2) and light penetration (by 60–160%). Elser et al. (1986) reported on the long-term response of liming Peter Lake between 1951 and 1976 through sampling conducted in 1984. The initial changes noted in 1951 persisted with little variation until at least 1984.

Aquatic Liming Practices After 1970

Liming to reduce acidity thought to be the result of acidic deposition has been practiced only rather recently, particularly in the United States. Table 1.3 describes the major liming programs conducted since 1970 in the United States, Canada, Sweden, Norway, and the United Kingdom. Similar programs have begun in other countries in western Europe. These liming programs were begun to reduce the acidity of surface waters believed to be affected by acidic deposition; to neutralize acidity due, in part, to unknown or undeterminable sources; and to provide reserve acid neutralizing capacity to protect surface waters from becoming acidic.

Only a brief summary of each major program is offered here; detailed information on the results of these liming programs is given in Chapter 3 and Appendices C and D.

United States

The number of lakes or ponds treated with alkaline materials probably number in the tens of thousands. Most of these were small ponds routinely treated to increase fish production. Only a small fraction of limed lakes and ponds have had careful liming and follow-up studies. About 150 surface waters in the United States have been limed and carefully evaluated since 1970. These projects were sponsored by the federal government (U.S. Fish and Wildlife Service, U.S. Environmental Protection Agency, U.S. Forest Service), state governments (Connecticut, Maryland, Massachusetts, Michigan, Minnesota, New York, Rhode Island, West Virginia, and Wisconsin), a private utility (Pennsylvania Power and Light), and not-for-profit organizations (Electric Power Research Institute and Living Lakes, Inc.). Most of the programs involved lake liming; only recently have a number of projects on streams been included. The distribution of liming sites by state is shown in Figure 1.1. The sites shown do not include liming conducted for treatment of acid mine drainage or for increasing

Table 1.3 Major Liming Activities Conducted Since 1970

Program	Location	Focus of Program or Project	Beginning Year	Reference
Electric Power Research Institute	New York (Adirondacks)	Experimental liming program for lakes in the Adirondack region of New York	1982	Porcella, 1989
New York State Dept. of Environmental Conservation	New York (Adirondacks)	Experimental and operational liming program for lakes in the Adirondack region of New York	1970s	Simonin et al., 1988
U.S. Fish and Wildlife Service Acid Precipitation Mitigation Program	Massachusetts, Minnesota, Tennessee, West Virginia	Research and demonstration of mitigation techniques on three streams and one lake (Minnesota)	1985	Schreiber, 1988
West Virginia Dept. of Natural Resources	West Virginia	Experimental and operational liming of selected impoundments and streams	1958	Zurbuch, 1963, 1984
Maryland Power Plant Research Program	Maryland	Evaluation of liming effects on anadromous fish in two coastal plain streams (co-funded with Living Lakes, Inc.)	1986	Janicki and Greening, 1988a,b

Organization	Location	Program	Year	Reference
Living Lakes, Inc.	Kentucky, Maryland, Masshachusetts, Michigan, New York, Pennsyl-vania, Rhode Island, West Virginia	Operational demonstration program for liming lakes, streams, and watersheds for restoration and protection of fisheries	1986	Brocksen and Emler, 1988
Ontario Ministry of the Environment; Ontario Ministry of Natural Resources	Sudbury, Canada	Experimental lake neutral-ization research program	1981	Booth et al., 1987; Dodge et al., 1988
Environment Canada	Nova Scotia	Neutralization of rivers to protect salmon fisheries (liming headwater lakes)	1982	Watt, 1986
Swedish Environmental Protection Board	Sweden	Operational and experimental liming of affected lakes and streamsthroughout Sweden	1972	Nyberg and Thornelof, 1988
Norwegian Liming Project	Norway	Operational and experimental liming of selected lakes and rivers in Norway	1979	Rosseland and Skogheim, 1984; Rosseland and Hindar, 1988

Table 1.3 Major Liming Activities Conducted Since 1970 (continued)

Program	Location	Focus of Program or Project	Beginning Year	Reference
Loch Fleet Project	Scotland, United Kingdom	Experimental watershed liming to treat Loch Fleet in southwest Scotland	1984	Brown et al., 1988
National Rivers Authority—Welsh Region (formerly Welsh Water)	Wales, United Kingdom	Three lakes and one watershed in central Wales	1984	Welsh Water, 1986 Underwood et al., 1987; Gee and Stoner, 1989

fish production in farmponds. About half the surface waters limed to date in the United States are in New York, specifically the Adirondack Park region. The major liming programs conducted in the United States since 1970 are described below.

Electric Power Research Institute

The Electric Power Research Institute (EPRI) has conducted aquatic liming in the Adirondack region of New York. The ongoing Lake Acidification Mitigation Program, which is sponsored by EPRI, focuses on research and demonstration of effective lake liming techniques (Porcella, 1989). Initial activities under the research program involved an evaluation of available data on liming efforts (Fraser et al., 1982; Fares and Kinsman, 1984). Actual ecosystem studies followed these evaluations. Three lakes were limed during 1985–1986. Extensive data were collected before, during, and immediately after liming to monitor changes in water quality, sediment chemistry, and aquatic biota. Sampling was continued in 1987 and 1988 to provide information on chemical and biological changes during reacidification. Results of liming treatments are given in Appendix C.

The *Canadian Journal of Fisheries and Aquatic Sciences* collected 13 papers that reported on the development of a dose and reacidification model for lake liming (Young et al., 1989; DePinto et al., 1989) and evaluation of how aquatic ecosystems respond to liming and later reacidification (Driscoll et al., 1989a,b); Staubitz and Zarriello, 1989; Gloss et al., 1989b; Roberts and Boylen, 1989; Schaffner, 1989; Fordham and Driscoll, 1989; Schofield et al., 1989; Evans, 1989; Bukaveckas, 1989). One of the papers provided an overview of ongoing and recently completed liming studies by EPRI (Porcella, 1989).

New York

The New York State Department of Environmental Conservation maintains a major operational lake liming program (Simonin et al., 1988). Kretser and Colquhoun (1984) summarized the New York liming program from 1964 to 1983. A total of 125 treatments were conducted on 56 lakes and ponds. Hydrated lime was used in 96 treatments made before 1976; agricultural limestone was

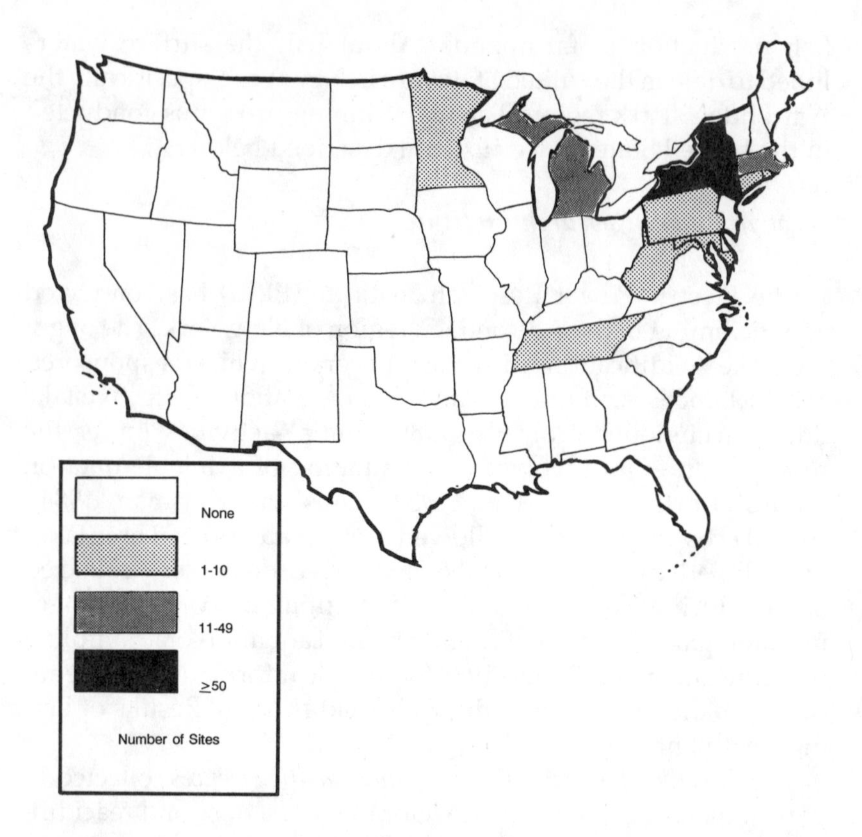

Figure 1.1 Distribution of sites in the United States in which lakes and streams were limed and evaluated. Sites are excluded where liming was conducted solely for the treatment of acid mine drainage or for increasing productivity of fish farm ponds.

used in 28 of the later treatments; and sodium carbonate was used in a more recent experiment. Some systems received multiple applications over time, e.g., Little Black Pond was treated seven times between 1959 and 1980.

From 1983 to 1985, 16 liming treatments were conducted by the state of New York. From 1986 to 1988 only five treatments were conducted due to higher priorities in other state environmental programs. During the entire period of 1959–1988, 67 waters were limed. A total of 142 treatments were conducted during this time, including various reapplications to the same lakes and ponds.

U.S. Fish and Wildlife Service

The U.S. Fish and Wildlife Service began the Acid Precipitation

Mitigation Program in 1985 for experimental liming of one lake in Minnesota and three streams—one each in West Virginia, Massachusetts, and Tennessee (Schreiber, 1988; Schreiber and Rago, 1984; Schreiber et al., 1988). The program focuses on research to improve the understanding of liming techniques and their effects on aquatic systems.

Thrush Lake in Minnesota was limed in May 1988. A slurry box was set up on a boat and 4.5 tonnes of powdered limestone were added to the lake over a 5-hr period to reach a target pH of 6.5 and an acid neutralizing capacity (ANC) of 100 μeq/L. Extensive monitoring of water chemistry and biota preceded and followed the treatment. The results are given in Appendix C.

Dogway Fork, a tributary stream of the Cranberry River in West Virginia, was selected for continuous liming; a self-feeding rotary drum system was used in which limestone aggregate was ground. The equipment was installed in July 1988 and liming was started in December 1988. The stream thereafter was limed continuously to maintain a pH of 6.5 and an ANC of 50 μeq/L. Water chemistry and biological monitoring began in 1986.

For the projects in Massachusetts and Tennessee, the collection of preliming data on water quality and fish was begun in 1986. A water-powered doser was installed at each site in early 1989 and powdered limestone treatments began in mid-1989. Effects of the treatments on water quality and biota have not yet been reported.

Another project jointly funded by the Service and the U.S. Environmental Protection Agency (EPA) involved evaluating 10 lakes in the Adirondack region of New York that were treated with limestone (Schofield et al., 1986; Gloss et al., 1989a). The project included an analysis of existing information from liming projects conducted by New York and private landowners associated with Adirondack League Club. The results are shown in Appendix C.

West Virginia

The West Virginia Department of Natural Resources has maintained a limited liming program that was begun the mid-1950s (Zurbuch, 1984; Fares and Kinsman, 1984). West Virginia in the late 1950s actively began pursuing promising techniques for neutralization of acidic streams and has continued to promote the rotary drum method of stream liming. Zurbuch (1984) described early attempts involving a rotary drum filled with limestone aggregate and placed on Otter Creek in the Monongahela National Forest. The installation on Otter Creek was operated until 1979. The initial success of this method in maintaining increased pH and alkalinity led to other rotary drum installations by groups such as Trout Unlimited and various coal mining companies. In the late 1970s the state began a study of a self-feeding system. A prototype was installed on Otter Creek in 1980. Results were encouraging enough to warrant continued experimentation, and in 1981 a cooperative project was begun with West Virginia University. The system was installed in 1983 on Otter Creek and performed well in a series of continuous runs. Further testing of the self-feeding system was conducted by the state as part of the Fish and Wildlife Service program.

Maryland

The Maryland Department of Natural Resources began liming two Coastal Plain streams in 1986, one on the Granny Finley Branch of the Chester River and another on Rock Branch of the Patuxent River (Greening et al., 1987; Janicki and Greening, 1988a,b). Wire mesh structures containing limestone were placed in each stream and water quality measurements were made on a semimonthly or monthly basis. In 1986, two other streams—Bacon Ridge Branch and Magothy Run—were selected as experimental streams to be treated during acidic episodes with limestone slurry applied by an automated doser. Extensive analyses of water quality and aquatic biota were conducted, but the unusually dry weather precluded full evaluation of the dosers. Because Magothy Run was strongly influenced by an impoundment immediately upstream, the doser was relocated to Mattawoman Creek in 1987. Operation of the dosers on these streams has continued. In December 1988, an experimental propane-powered doser was installed and tested on Stocketts Run, a tributary of the Patuxent River (Greening et al., in press).

Living Lakes, Inc.

Living Lakes, Inc. (LLI), a not-for-profit fish restoration demonstration program, was formed in 1986 with funding from private sources to demonstrate the effectiveness of liming (Brocksen and Emler, 1988). Through 1989, seven liming techniques have been evaluated on 37 lakes and 11 streams in 6 states: Kentucky, Maryland, Massachusetts, Michigan, New York, Pennsylvania, Rhode Island, and West Virginia (Olem et al., 1990). The program includes ongoing water quality analyses for the lakes and streams treated and, if necessary, retreatment. Recently, LLI began investigating the treatment of whole or partial watersheds (Brocksen et al., 1988). These projects, cosponsored by other organizations, are being conducted in the Black Forest in Germany and in the Woods Lake watershed in the Adirondacks.

Canada

In Canada, early liming experiments involved liming four lakes in the Sudbury area of Ontario (Scheider et al., 1975; Scheider and Dillon, 1976; Yan et al., 1977a,b, 1979; Yan and Stokes, 1978; Dillon et al., 1979; Goodchild and Hamilton, 1983; Scheider and Brydges, 1984; Yan and Dillon, 1984). Later experiments near Sudbury, Ontario, were conducted on Bowland Lake and Trout Lake (Molot et al., 1984, 1986; Molot, 1986; Booth et al., 1986, 1987; Dodge et al., 1988). Liming of lakes and rivers has also been evaluated in Nova Scotia (Watt et al., 1984; White et al., 1984; Watt, 1986).

Ontario

In 1981, the Ontario Ministry of the Environment and the Ontario Ministry of Natural Resources began a 5-year investigation of the feasibility of neutralization as a technique to rehabilitate acidified lakes and to protect acid-sensitive lakes (Dodge et al., 1988).

Bowland Lake and Trout Lake, both in the Sudbury area, were limed in 1983 and 1984, respectively. Another lake was selected as a nonacidic control lake for bioassay experiments. A final technical report on the first 5 years of study was issued by Booth et al. (1987). The results are shown in Appendix D.

Nova Scotia

Liming of lakes and salmon rivers believed to be affected by acidic deposition has been considered by Environment Canada in Nova Scotia (Watt et al., 1984; White et al., 1984; Watt, 1986). In 1981, the Canadian Department of Fisheries and Oceans carried out an experimental neutralization (by liming) of Sandy Lake, a headwater lake in the Sackville River system near the town of Bedford, Nova Scotia. The purpose of the liming was to determine whether neutralization of headwater lakes could be used to protect and restore Atlantic salmon habitat in the river system downstream from the lake. In July and August 1981, 135 tonnes of powdered limestone were added to a headwater lake by boat. The lake liming increased the pH in the river downstream, although neutral conditions were not maintained because of the high flushing rates of the lakes and heavy winter rains.

Sweden

Sweden has the largest and most comprehensive aquatic liming program in the world (Nyberg and Thornelof, 1988). Technical and administrative results from the program have been given in a number of publications (Bengtsson et al., 1979, 1980; Rodhe, 1981; Swedish Ministry of Agriculture Environment Committee, 1982; Nyberg, 1984; Dickson, 1985; Fraser et al., 1985; Hasselrot and Hultberg, 1984; Lessmark, 1987; Lessmark and Thornelof, 1986; Dilworth, 1988; Krogerstrom, 1988a,b; Melander, 1988; Sverdrup and Warvinge, 1988a,b; Thunberg, 1988; Nyberg, 1988, 1989).

Liming of acidified waters in Sweden, with the aid of government subsidies, was performed for a trial period from 1977 to 1982, during which about 1100 lakes were treated. On the basis of the success of the treatments, the Swedish government began liming on an operational basis, expanding the subsidy program. Additionally, an extensive research program was begun on 96 lakes to study the feasibility of using different liming techniques, fertilization, intensive fishing, and selenium treatment to reduce the accumulation of mercury in fish (Nyberg and Thornelof, 1988).

Through 1988, about 5000 surface waters were treated in the operational liming program conducted by the Swedish government (Nyberg, 1989). The lakes have been relatively small, having a median surface area of 22 ha and a maximum depth of 9 m (Nyberg and Thornelof, 1988). Retention times were less than 1 year in most of the treated lakes, but exceeded 5 years in 60 of them. The median pH before treatment was 5.7. All lakes were classified as oligotrophic before treatment and normally were inhabited by no more than six fish species. During the operational phase, nearly all of the waters were treated with powdered limestone. Over half the treatments were conducted by applying slurried limestone directly from boats. About 20% of the treatments were conducted by using trucks and tractors on ice-covered lakes or in watersheds, and in the rest of the liming applications, automatic dosers, diversion wells, and helicopters were used. Most of the liming operations were conducted on lakes; considerably fewer streams were treated.

The most comprehensive individual study of lake liming in Sweden was conducted at Lake Gardsjon (Dickson, 1985, 1988). Appendix D summarizes the results of a detailed evaluation of the physical, chemical, and biological effects of liming over an 8-year period.

Norway

Although surface water acidification problems in Norway are similar in many ways to those in Sweden, Norway has begun a somewhat smaller government-supported liming program (Baalsrud, 1988; Hindar and Rosseland, 1988; Kleiven et al., 1989a,b; Langvatn, 1989). The Norway program focuses efforts on waters in the more acidified southern region and includes both operational and experimental liming components. Government subsidies have resulted in the treatment of several hundred lakes and a smaller number of streams, rivers, and watersheds (B.O. Rosseland,personal communication, 1990). The operational program is second only to Sweden in the number of surface waters limed.

The liming of Lake Hovvatn, which began in 1980, has been the most comprehensive of the Norwegian experimental liming projects (Wright, 1984, 1985; Kleiven et al., 1989). Results of the liming Lake Hovvatn and the watershed liming at Tjonnstrond are given in Appendix D.

Recent operation-oriented projects in Norway include the River Audna, the catchment Vegar, the River Vikedalselva, and two smaller lakes in southern Norway (Hindar and Rosseland, 1988; Kleiven et al., 1989a,b; Langvatn, 1989). The River Audna has been limed continuously since 1985, primarily to improve the Atlantic salmon fishery. An upstream lake was limed in 1985 and two continuous limestone dosing devices have been used since then to neutralize both lake and river water throughout the year. The catchment Vegar includes a 60-ha lake limed in 1985 and 1987, and a salmon river limed in 1987. The River Vikedalselva in southern Norway is being limed to prevent acute mortality of Atlantic salmon smolts. One large doser has added limestone to the river since 1987. The doser is operated continuously in the spring and is the only project with the dose controlled automatically according to the pH downstream of the liming. The two smaller lakes described by Hindar and Rosseland (1988) included one limed in 1985 with a dolomite powder and the other limed with different limestone powders periodically since 1981.

United Kingdom

The focus of liming research in the United Kingdom has been largely experimental. In Scotland, the emphasis has been on two lakes, Loch Fleet and Loch Dee. Lake and watershed liming has recently been conducted in central Wales; treatment of three lakes began in 1984 and liming of the watershed of a large reservoir in 1987.

Scotland

The most comprehensive liming project in Scotland has been the Loch Fleet Project, an experimental program to test the suitability of lake, stream, and watershed liming for improving the water chemistry and trout population of the lake (Brown et al., 1988). The project was begun in 1984, with an initial phase between 1984 and 1986 that focused on collecting baseline information in anticipation of future treatments (Howells, 1987). Howells (1986) and Howells et al. (1986) reported on the results of the first phase of the study, and Brown et al. (1988) reported the results of the first set of treatments applied in spring 1986. Selected locations in three subwatersheds were treated and monitoring was conducted at the lake inlets from the subwatersheds and at the lake outlet. Results are given in Appendix D.

Burns et al. (1984) described preliminary investigations at Loch Dee in southwest Scotland. The 100-ha lake was considered an appropriate candidate for limestone application. Tervet and Harriman (1988) described the liming of the Loch Dee watershed on five occasions between 1980 and 1983. A total of 255 tonnes of limestone powder was applied. Much of the material added during the first 2 years was absorbed by soils and sediments, and was not effective in substantially increasing the pH of the lake. Lake treatments were more effective because limestone was applied to major water pathways and headwater streams.

Wales

Liming studies in Wales were reported by Welsh Water (1986), Underwood et al. (1987), and Gee and Stoner (1989). In 1984, Welsh Water (1986) used lake liming to neutralize acidic conditions in Llyn Pendam, a lake in central Wales. Two lakes in central Wales, Llyn Berwyn and Llyn Hir, were treated with limestone powder in 1985. Water quality was monitored after liming and was reported by Underwood et al. (1987) to be suitable for the survival of stocked fish species. Brown trout were stocked in the lakes 1 month after limestone addition and their survival was confirmed during later monitoring (Welsh Water, 1986).

Gee and Stoner (1989) described the application of limestone powder to the watershed at Llyn Brianne in 1987. Llyn Brianne is a relatively large reservoir in Wales with a surface area of 215 ha. The effects of the treatments on water quality and biota have not yet been reported.

2

Mitigation Strategies

"Whoever wants to reach a distant goal must take many small steps."
Helmut Schmidt

INTRODUCTION

Several methods are available for mitigation of acidic surface waters, most of which involve the addition of base materials to neutralize the acidity of surface waters, sediments, and soils. *Liming* is the generic term for the addition of any base material, whether it be ground limestone (the most common material used) or other alkaline chemicals such as soda ash (sodium carbonate). Other techniques, such as nutrient addition, stocking of acid-tolerant fish strains, and pumping of alkaline water, have been proposed for mitigation of adverse effects of acidic surface waters, but have not been adequately tested. The various methods (Table 2.1), including those still in the research stage, are discussed here.

Liming strategies include the addition of limestone or other neutralizing materials directly to the lake or stream, injecting alkaline materials into lake sediments, adding base material to wetlands or watersheds, and placing filters or barriers of base material in streams. Each strategy can vary, depending on the type of alkaline material available for neutralization and the means of application. The application method selected depends on whether the surface water to be treated is a lake or a stream. Treatment methods for lakes can also differ depending on the lake's size, flushing time, access, and whether the lake is managed to provide a "put-and-take" fishery or a self-sustaining population of fish. Stream treatment techniques can differ, depending on the size of the system, availability of an electric power source, access, and management goals. Cost is also a very important factor in determining which mitigation technique is selected.

Table 2.1 Methods Available for Mitigation of Acidic Surface Waters

Method	Description	Relative Use
Liming surface waters	Application of base materials to lakes and streams	Common
Liming lake sediments	Application of base materials to sediments by injection	Uncommon
Liming wetlands and watersheds	Application of base materials to wetland or watershed areas upstream from acidified lakes or streams	Occasional
Nutrient addition to surface waters	Application of phosphorus or organic carbon to lakes	Uncommon
Fish stocking	Stocking of lakes with acid-tolerant or acid-resistant fish strains	Occasional
Pumping of alkaline water	Pumping of water into lakes or streams from a source of relatively more alkaline water, such as groundwater	Uncommon

TYPES OF AQUATIC RECEPTORS AND TREATMENTS

The receptors considered here include lentic systems (lakes, reservoirs, and ponds), lotic systems (streams and rivers), and watershed areas. Strictly speaking, watershed areas are usually part of the treatment rather than the target for neutralization. Different treatment strategies are generally required within and between receptors.

Lakes have been the most commonly treated receptors to date. The Swedish government liming program, involving more than 5000 surface waters, provides an example of the distribution of receptors. Figure 2.1 shows the relative distribution of mitigative treatments among different receptors in the Swedish program during 1983–1987 (Nyberg and Thornelof, 1988). About 70% of the quantity of limestone used in 1987 (149,000 tonnes) was applied directly to the lake surface and the rest was evenly divided between streams and wetlands (Nyberg and Thornelof, 1988; Svensson and Bratt, 1989).

Lentic Systems

Lentic systems have been the receptor most widely treated for

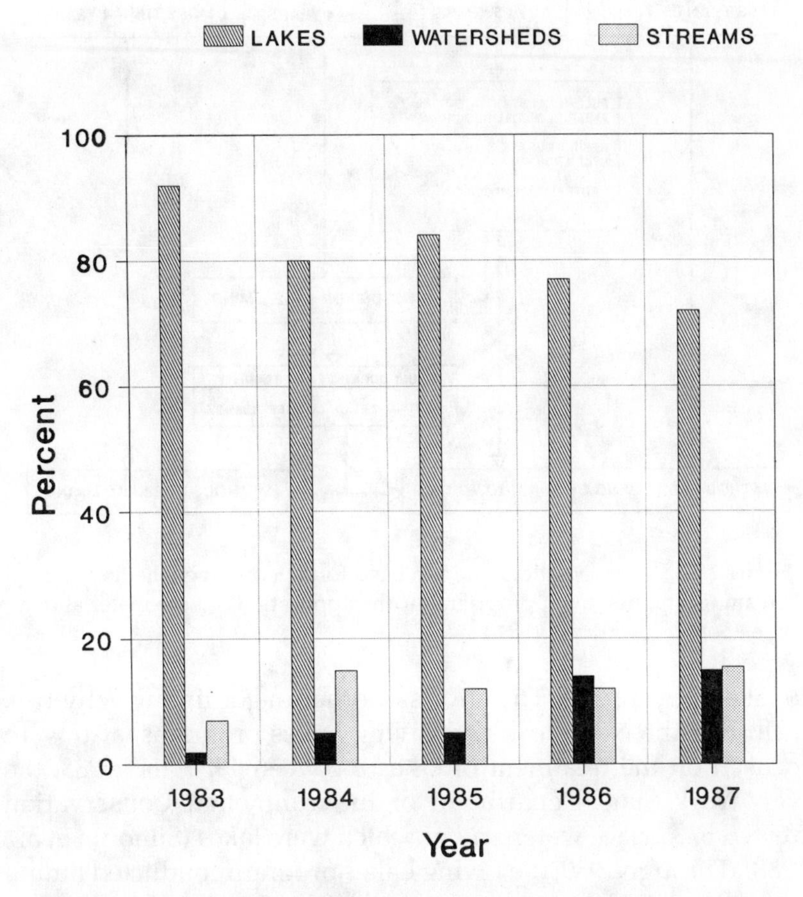

Figure 2.1 Relative distribution of limestone addition in Sweden for different receptors. (Source: Nyberg and Thornelof, 1988.)

mitigation of acidic conditions, primarily because cost-effective treatment is possible in many lakes with a single application. This may last for several years if the lake's water retention time is long. Flowing waters, on the other hand, generally require the installation of permanent structures for continuous treatment. For lakes with extremely short water retention times, the application techniques described later for running waters and watersheds may be more appropriate than those described for lakes.

Because the technology for continuous treatment of flowing waters has been advanced only recently, the emphasis in mitigation programs has been on lakes. During the 5 years of the Swedish operational liming program, 70–90% of the limestone used each year for the mitigation of acidic conditions was applied

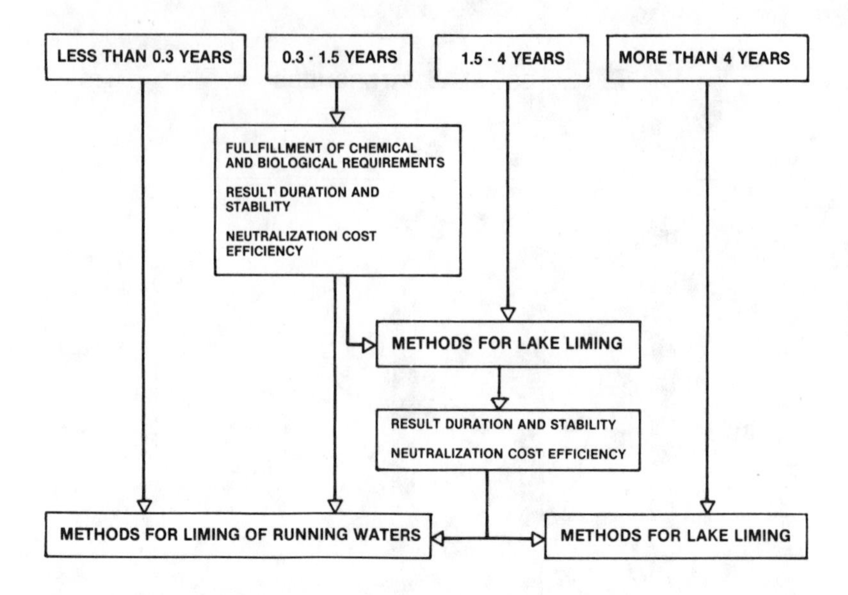

Figure 2.2 Water retention time as a basis for determining whether lake or stream techniques are appropriate for the application of base material to a lake. (Source: Sverdrup, 1984.)

to lakes (Figure 2.1). The efforts of other major liming activities, including the New York and Living Lakes programs, also were focused on the treatment of lakes. Between 1959 and 1988, the New York State Department of Environmental Conservation treated 67 surface waters, all of which were lakes (Simonin et al., 1988). Through 1989, the Living Lakes program conducted liming activities on 37 lakes and 11 streams (Olem et al., 1990).

Although treatment of lakes by direct application of base materials to the water surface is the most common mitigation technique used to date, treatment of the watershed to protect lakes and flowing waters has been receiving increased attention in recent years (Appendix D).

Inasmuch as single application to the surface may not be effective in lakes with rapid flushing rates, Sverdrup (1984) developed a strategy to determine whether a lake should be treated by using methods suited for flowing water (Figure 2.2). Stream techniques were considered more appropriate for treating lakes with retention times of less than 0.3 year. Review of other factors was considered necessary before selecting methods for stream or for lakes with retention times between 0.3 and 4 years. Only lake

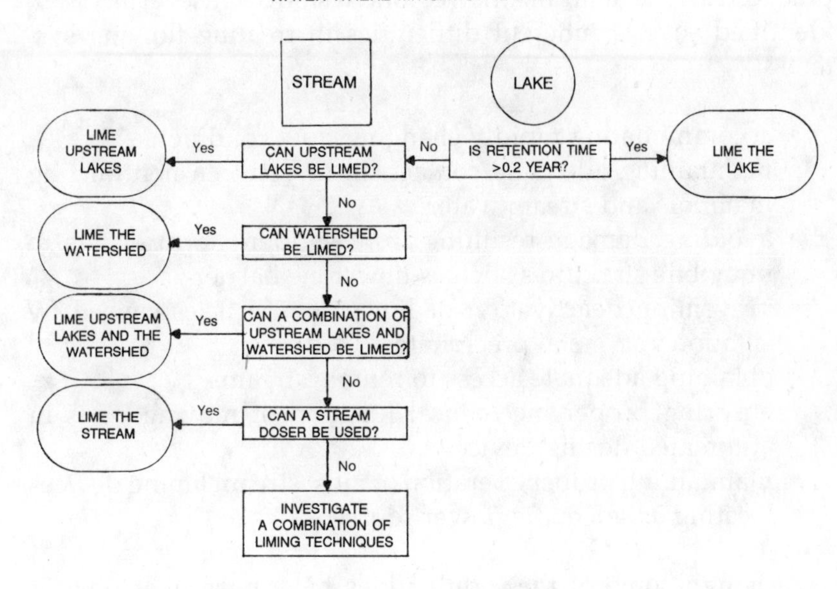

Figure 2.3 Decision strategy presented in the Swedish Liming Handbook for determining recommended liming methods. (Source: Swedish Environmental Protection Board, 1988.)

liming techniques were considered appropriate for the treatment of lakes with retention times greater than 4 years.

Another decision strategy was presented in the Swedish Liming Handbook (Figure 2.3; Swedish Environmental Protection Board, 1988). Direct liming of a lake was not recommended for lakes with retention times less than 0.2 year. In situations where lake retention time is less than 0.2 year, it is recommended that other factors be reviewed to determine whether watershed applications, stream techniques, or combination treatments are appropriate.

Lotic Systems

Few lotic systems (streams and rivers) thought to be affected by acidic deposition have been treated for mitigation of acidic conditions, primarily because experience with mechanical dosing devices (the technology that has been most successful) has been limited. Also, the liming of upstream lakes or watersheds has been practiced for the neutralization of lotic systems. Fraser et al. (1985) identified several inherent difficulties in treating flowing systems:

- accommodating rapidly changing flow conditions
- maintaining adequate contact times between neutralizing materials and stream water
- avoiding damage resulting from sudden storm events to immobile structures such as limestone barriers
- preventing deactivation of limestone particles caused by siltation and metal precipitates
- obtaining adequate access to remote streams
- ensuring proper mechanical function of moving parts in automated dosing devices
- maintaining proper operation of most stream liming devices
- locating an adequate power source

Although some of these difficulties have been overcome in recent years with advances in automatic dosing devices, methods for direct treatment of lotic systems remain generally far less developed than those for lentic systems.

Liming treatments applied to streams as part of the Swedish program have nearly doubled from about 8% of the total limestone used in 1983 to 15% in 1987 (Figure 2.1). Other programs also have begun liming treatments on streams as technology has advanced. Because of the interest in restoring and protecting stream fisheries in West Virginia, Pennsylvania, and Maryland, liming activities in these states have focused almost exclusively on streams (Zurbuch, 1984; Janicki and Greening, 1988a,b; Arnold et al., 1988).

Streams and rivers can be treated indirectly by adding base materials to an upstream lake, a method that has been successfully practiced in Canada and Sweden (Watt et al., 1984; Edman and Fleischer, 1980). Watt (1986) estimated that 70% of the acidic rivers in Nova Scotia could be treated by liming headwater lakes. Sites amenable to this type of mitigation strategy in other areas have not been estimated.

Watersheds

Soil liming, i.e., the addition of neutralizing material to portions of the watershed, is considered to be a viable alternative to the direct addition of base materials to surface waters. *Soils* are used here in a broad sense to mean areas other than lakes or streams, and to include dry soils and wetlands. Experience with watershed liming has indicated that it is important to apply the base material to major water pathways. This practice avoids treatment of the entire watershed and reduces the amount of base material required.

The most successful projects have been those where limestone was applied directly to discharge areas (portions of the watershed where groundwater rises to the surface during baseflow or stormflow conditions). The usefulness of treatment to merely recharge areas (i.e., dry soils) has been limited; on the basis of the results of 52 soil liming projects, Lessmark (1987) reported discouraging results for treatments where limestone was applied only to recharge areas. The surface waters were effectively neutralized only when liming included discharge areas.

The principal advantage of watershed liming is that the effects of this type of treatment are more sustained relative to surface water treatments (Warfvinge and Sverdrup, 1988a,b; 1989; Sverdrup and Warfvinge, 1988b; Brown et al., 1988; Rosseland and Hindar, 1988). This advantage may greatly influence the total cost as well as cost effectiveness) of watershed treatment compared to that of other approaches. Other possible advantages are the neutralization of intragravel areas that may occur after watershed liming, which may protect the spawning grounds of certain species, and the mitigation of acidic conditions in streams and the littoral zones of lakes during storm episodes.

Although the practice has been relatively uncommon, soil liming for mitigation of acidic surface waters has increased in recent years. For example, about 2% of the total limestone used in mitigative treatments in Sweden was applied to soils in 1983, the first year of the Swedish program (Figure 2.2). The liming of wetlands or watersheds has increased in Sweden; by 1987 about 15% of the total limestone used was applied to wetlands.

Until recently, procedures for determining dose calculation and surface water reacidification rates did not exist for soil liming (Sverdrup and Warfvinge, 1988a,b). Current procedures remain somewhat limited, however, providing only an approximation of the required dose because of the extremely complex acid and base chemical reactions in soil systems (Warfvinge and Sverdrup,

1988a,b). The dose depends heavily on the choice of locations for treatment in the watershed. The dose required is lower if areas of high discharge are located for the application of base material. A large portion of the base material used was unavailable for lake neutralization when whole watersheds or subwatersheds were treated (Tervet and Harriman, 1988: Brown, 1988).

The direct application of bases to lakes with water retention times of less than 6 months is ineffective because neutralization tends to be temporary and lakes can often reacidify after snowmelt or after a single heavy rain. Soil liming may be particularly applicable in these situations, because it potentially reduces both the severity of episodic acidic conditions and the leaching of toxic aluminum from the edaphic to the aquatic environment. Aluminum leaching has been shown to sometimes result in biological problems after the direct liming of surface waters (Nyberg, 1984). In some situations, however (such as northern regions with heavy winter snowpack and frozen ground), watershed liming may not reduce the severity of episodic acidification; the lack of effectiveness of this technique may result from the limited interaction between water derived from storms or snowmelt and neutralized soils (Fraser et al., 1985). This phenomenon has not been observed in the field, however. After watershed liming at Loch Fleet in Scotland (Brown et al. 1988) and Lake Tjonnstrond in Norway (Rosseland and Hindar, 1988), pH and calcium increased after storms. Limestone in soils apparently neutralized precipitation, allowing neutralized water to reach the lake.

Combination Treatments

A combination of direct addition of a base to a receptor and to the soil has been suggested to provide both immediate and sustained results. Combination treatments are commonly used in the Swedish liming program (Swedish Environmental Protection Board, 1988). Dose calculation methods, however, are largely undeveloped and reacidification rates are largely unknown.

LIMING TECHNIQUES FOR MITIGATION

In this section available neutralizing agents and the difference between lakes and stream neutralization techniques are described;

in addition, whole, lake neutralization, neutralization of portions of lakes, and liming wetlands and watersheds are further distinguished.

Neutralizing Agents

Many alkaline materials have been used or proposed for neutralizing acidic waters (Grahn and Hultberg, 1975; Fraser et al., 1982; Fraser and Britt, 1982; Davison and House, 1988). These materials can be divided into four main groups: carbonates, oxides, hydroxides, and silicates (Table 2.2). The theoretical neutralization equivalents of some of them have been compared (Table 2.3). Theoretical neutralization equivalents are expressed relative to an assigned value of 100% for pure calcium carbonate. The higher the number, the less material it would take for neutralization per unit weight and unit acidity. For example, calcined lime and calcined dolomite are more efficient neutralizers than calcium carbonate; sodium carbonate is less efficient. However, other facts must be considered, such as cost, availability of materials, application method, dissolution efficiency, safety, and effects of materials on water chemistry and biota.

Waste-containing chemicals in one or more of these major groups of materials have also been proposed for neutralizing waters, but their use in aquatic liming situations has been rare. These materials include industrial by-products and wastes, such as cement plant bypass dust, sludge from water-softening treatment facilities, and fly ash (Lovell, 1973). Whether manufactured products or waste, these materials have different characteristics that make them suitable for particular neutralization situations. The materials may be unsuitable because they tend to contain high concentrations of trace metals and other impurities. Waste materials not readily available commercially are not further described, because their use in aquatic liming situations is extremely limited.

Carbonates

Commonly used carbonates are calcium carbonate (commonly referred to as *high-calcium limestone* or *calcite*), calcium-magnesium carbonate (dolomite), sodium carbonate (soda ash), and sodium bicarbonate (baking soda).

Table 2.2 Characteristics of Neutralizing Materials

Group	Formula	Common Name	Relative Dissolution Rate	Potential for Excessive pH Values if Dosages are Miscalculated	Relative Availability	Relative Cost
Carbonates	$CaCO_3$	Aglime, high calcium limestone, calcite, calcium carbonate	Slow	None	High	Low
	$CaCO_3$-$MgCO_3$	Dolomitic limestone, dolomite, calcium-magnesium carbonate	Very slow	None	High	Low
	Na_2CO_3	Sodium carbonate, soda ash, soda	Instant	High	High	High
	$NaHCO_3$	Baking soda, sodium bicarbonate	Very rapid	Moderate	High	High
Hydroxides	$Ca(OH)_2$	Slaked lime, hydrated lime, high calcium hydrated lime	Rapid	High	High	Med–high
	$Ca(OH)_2$-MgO or $Ca(OH)_2$-$Mg(OH)_2$	Dolomitic hydrate, dolomitic hydrated lime	Rapid	High	Medium	Med–high
	$NaOH$	Caustic soda, lye, sodium hydroxide	Instant	High	High	High
Oxides	CaO	Calcined lime, quicklime, high calcium quicklime, calcium oxide	Rapid	High	Medium	Medium
	CaO-MgO	Calcined dolomite, dolomitic quicklime, calcium-magnesium oxide	Rapid	High	Medium	Medium
Silicates	$(CaO)_3$-SiO_2	Basic slag	Slow	Moderate	Low	Low
	$(MgO)_2$-SiO_2	Olivine	Very slow	None	Low	Low

Table 2.3 Theoretical Neutralization Equivalents of Selected Materials

Common Name	Formula	Formula Weight	Theoretical Neutralization Equivalent (%)[a]
Limestone	$CaCO_3$	100.09	100
Dolomite	$CaCO_3$-$MgCO_3$	184.42	109
Sodium carbonate	Na_2CO_3	106.00	94
Sodium bicarbonate	$NaHCO_3$	84.00	119
Calcined lime	CaO	56.08	179
Calcined dolomite	CaO-MgO	96.40	207
Hydrated lime	$Ca(OH)_2$	74.10	135
Dolomitic hydrate	$Ca(OH)_2$-MgO	114.42	175
Pressure dolomitic hydrate	$Ca(OH)_2$-$Mg(OH)_2$	132.44	151
Caustic soda	NaOH	40.01	125

[a] Relative neutralization equivalent if pure $CaCO_3$ is assigned a value of 100%.

Calcium Carbonate

Calcium carbonate ($CaCO_3$), herein referred to simply as *limestone,* is the neutralizing agent most commonly used for mildly acidic surface waters (Fraser and Britt, 1982; U.S. Fish and Wildlife Service, 1982; Zengerie and Allan, 1987; Huckabee et al., 1989; Porcella et al., 1989). The U.S. Fish anad Wildlife Service program for evaluating surface water liming (Acid Precipitation Mitigation Program) considered only ground limestone for use at the research sites (Fraser et al., 1985). The cost of limestone is low, because it is used extensively for agricultural liming, and its availability is high. It is a naturally occurring mineral product and a component of the surface water buffering system. Its quality and reactivity vary widely with origin (Lovell, 1973).

Other advantages cited for the use of limestone are its relatively slow dissolution rate, which usually results in a gradual increase in pH; the low likelihood of pH increases to excessive levels if dosages are miscalculated (biological stress resulting from pH > 9); and its low causticity, which reduces the hazards of handling and application (Grahn and Hultberg, 1975; Schneider et al., 1975; Edzwald and DePinto, 1978; Bengtsson et al., 1980; Hultberg and Andersson, 1982; Theiss, 1981; Boyd, 1982a; Sverdrup, 1982a,b; Melander, 1988; Porcella, 1989). In addition, the high calcium content of limestone has been reported to provide additional

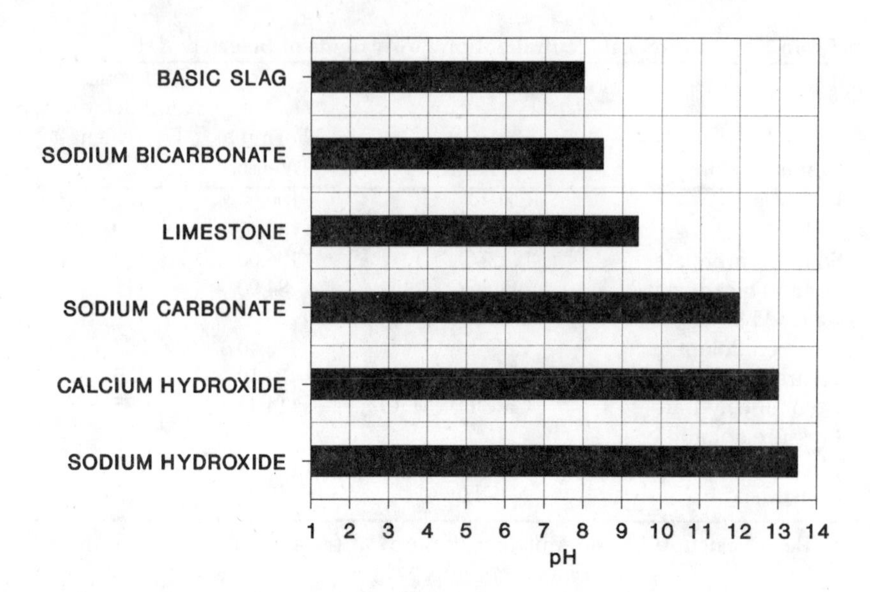

Figure 2.4 Maximum pH obtainable in the field where certain neutralizing materials are used.

protection for aquatic organisms (Brown, 1982a,b; Hunn et al., 1985; Rosseland and Skogheim, 1984).

Stumm and Morgan (1981) reported the maximum pH attainable when an aqueous system is in equilibrium with $CaCO_3$ and atmospheric CO_2 is 8.3. In practice, maximum pH values for limestone are generally lower after equilibrium is reached. Equilibrium conditions, however, may not occur immediately and pH values above 9 have been reported for as long as 2 weeks after treatment (Fordham and Driscoll, 1989). Figure 2.4 illustrates the pH ranges of waters treated with various neutralizing materials. Limestone treatment generally results in maximum attainable pH values of 7–8.

The dissolution rate of limestone is a function of particle size, chemical properties of the stone, chemistry of the receiving water body, turbulence of the receiving water body, water temperature, and depth of water. Adams and Brocksen (1988) reported that particle size is the single most important factor in the dissolution of high-calcium limestone. This observation was also made by Sverdrup and co-workers (Sverdrup, 1982a, 1982b, 1984, 1985; Sverdrup and Bjerle, 1982, 1983; Sverdrup and Warfvinge, 1988a,b). Boyd (1982b) reported that the particle size and neutralizing values of limestone can provide an estimate of its overall effectiveness in neutralizing acidity in surface waters.

Finely ground stone results in increased dissolution because of the larger surface area available for reaction. Consequently, powdered limestone is often used. Powdered limestone with larger particle sizes (commonly referred to as *agricultural limestone*) is not normally recommended for lake treatment because of its poor dissolution characteristics in moderately acidic surface waters. Larger crushed stone can be used in rotary drums or limestone barriers. More porous stone also allows greater dissolution. The particle size of limestone materials is commonly measured with either wet sieving or Coulter counter techniques.

The chemical properties of the stone also are important. Limestone can vary in calcium and magnesium content, may contain excessive concentrations of trace elements such as lead, and sometimes includes large amounts of insoluble material.

The chemical conditions of the receiving water also influence limestone dissolution. Dissolution rates are greatest in water with low pH, low dissolved inorganic carbon, and low calcium—all characteristics of waters acidified by coal mining practices. Finely ground limestone can compensate for slower dissolution rates, especially in surface waters with pH values higher than 5.7 (Sverdrup, 1982a). Water temperature is another important factor; dissolution rates increase with increasing water temperature (Sverdrup, 1985). It has been noted, however, that the enhanced dissolution rate at higher temperatures may not increase net dissolution, because limestone particles sink faster, thus increasing the likelihood of significant amounts of material settling to the bottom of a lake.

Organic carbon, iron, and other materials in the water sometimes coat the surfaces of limestone and inactivate the neutralization process. This has been a common occurrence when crushed limestone has been placed in streams (Pearson and McDonnell, 1975a,b; Bernhoff, 1979; Driscoll et al., 1982). The plating of exposed limestone surfaces with insoluble metal hydroxides and carbonate precipitates makes it difficult to predict whether the material that does not dissolve immediately (within several weeks) eventually becomes inactive. Limestone settled on the lake bottom, however, has been shown to continue to dissolve over several years. Alenas (1986) reported continued dissolution of limestone over a 7-year period in Lake Lilla Harsjon, possibly due to biological activity and CO_2 in upwelling groundwater.

Physical characteristics of the water are important determinants in the dissolution of limestone. More turbulent and deeper waters allow more effective dissolution and reduce the likelihood of significant amounts of material settling to the bottom of a body of water. Warmer water dissolves limestone more rapidly. Lake stratification characteristics also affect limestone dissolution.

Calcium-Magnesium Carbonate

Dolomitic limestone ($CaCO_3$-$MgCO_3$), although lower in calcium content than limestone (the magnesium content exceeds 35%), is also readily available. The neutralization equivalents of dolomite is slightly higher than that of limestone (Lewis and Boynton, 1976). Limestone containing more than 10% $MgCO_3$, however, dissolves too slowly in acidic systems to be of practical use in the neutralization of most acidified surface waters (Lewis and Boynton, 1976; Bernhoff, 1979; Sverdrup, 1985).

Sodium Carbonate

Sodium carbonate (Na_2CO_3, commonly referred to as *soda ash*) has been used in a few applications for surface water neutralization (Stoclet, 1979; Lindmark, 1982, 1985; Smith, 1985). Since Na_2CO_3 is a slightly soluble base, its use results in only minor losses from incomplete dissolution and produces less extreme fluctuation in pH than do oxides and hydroxides. In practice, the maximum pH attainable with Na_2CO_3 is pH 11 or 12. Because soda ash is used in many industrial applications and in water and wastewater treatment, it is readily available. Although it is reported to be a more effective neutralizing agent than $CaCO_3$ on an equivalent weight basis (Table 2.3), its cost is about five times that of $CaCO_3$. Calcium-based neutralizing agents also provide calcium ions that have been shown to be beneficial for the growth and survival of aquatic organisms (Brown, 1982a,b; Rosseland and Skogheim, 1984; Hunn et al., 1985). Sodium is a more effective exchanger in cation exchange reactions with lake sediments than calcium, and it has been reported to be particularly effective for lake sediment injection methods of neutralization (Stoclet, 1979; Lindmark, 1982, 1985). This method is discussed later.

Sodium Bicarbonate

Sodium bicarbonate ($NaHCO_3$, commonly referred to as *baking soda*) has rarely been used in the neutralization of acidified surface waters, but it has been used in the treatment of drinking water and wastewater (Church and Dwight, 1989). Its relative availability, like that of soda ash, is high, but its cost also is high compared to that of calcium-based materials. Unlike soda ash or sodium hydroxide, baking soda yields a maximum pH of only about 8 in the field; thus, unintended overneutralization to pH values that may be detrimental to aquatic life is impossible. Baking soda is also safe and easy to handle, and dissolves rapidly. It is reported to be more effective than $CaCO_3$ as a neutralizing agent, on an equivalent weight basis (Table 2.3). Because the cost of baking soda is relatively high, its use has been reported only once in the United States: Wolf Pond (20 ha) in northeastern Franklin County in the Adirondack region of New York was treated in 1987 (Siegfried et al., 1987).

Oxides

The oxide materials used for the neutralization of water are calcium oxide (CaO or quicklime) and calcium-magnesium oxide (CaO-MgO or dolomitic quicklime). Their use is much more common in drinking water, sewage, and industrial wastewater than in lakes and streams.

Calcium Oxide

Quicklime (also referred to as *high-calcium quicklime* and *calcined lime*) is a product resulting from the calcination of limestone. Calcination occurs when limestone is "burned" in a kiln at a temperature of about 1000°C. In this process, carbon dioxide gas is driven off and CaO is produced. Quicklime, or high calcium quicklime, is defined here as CaO containing less than 5% MgO.

Quicklime dissolves rapidly in water and can increase pH values to about 13. Because it is highly caustic and difficult to handle, store, and apply safely, it is rarely used in neutralizing natural waters. Exposure to moisture can result in the formation of water-soluble carbonates. Quicklime is generally not as readily available as hydrated lime; its cost is somewhat lower than that of hydrated lime but higher than that of limestone.

Quicklime is available in a number of standard particle diameters, most commonly in ground and pulverized forms (National

Lime Association, 1981). Most ground quicklime passes through a #8 sieve, and 40–60% passes through a #100 sieve. Pulverized quicklime results from more intense grinding and most of it passes through a #20 sieve, and 85–95% through a #100 sieve.

Calcium-Magnesium Oxide

Dolomitic quicklime (CaO-MgO, with a MgO content greater than 35%), although lower in calcium content, is also readily available. Its neutralizing value is similar to that of high-calcium quicklime. The use of dolomitic quicklime offers no particular advantages over high-calcium quicklime for neutralizing acidic surface waters.

Hydroxides

Hydroxides that have been used for neutralization of water are calcium hydroxide (commonly referred to as *hydrated lime* or *slaked lime*), calcium-magnesium hydroxide (dolomitic hydrated lime), and sodium hydroxide (caustic soda). Their use is much more common in drinking water, sewage, and industrial wastewater treatment than in lake and stream applications. Calcium hydroxide was the primary neutralization agent used in liming treatments conducted in the United States before 1970.

Calcium Hydroxide

Hydrated lime [$Ca(OH)_2$] is a product manufactured by adding CaO to water at a ratio of about 1:2. The hydration process results in a product that is less likely to react chemically with moisture in the atmosphere, and thus the material is less caustic than quicklime and somewhat safer to handle. Calcium hydroxide is, however, much more caustic than calcium carbonate. Relative to $CaCO_3$, $Ca(OH)_2$ is more reactive and water soluble (Lewis and Boynton, 1976; Boyd, 1982b; Swedish Ministry of Agriculture Environment Committee, 1982). Consequently, less $Ca(OH)_2$ is required to produce the same ANC in a body of water. The higher dissolution rate, however, is often considered a detrimental attribute: when $Ca(OH)_2$ is directly applied to surface waters, it rapidly increases pH, potentially inducing pH shock in aquatic biota (Simonin et al., 1988; Driscoll et al., 1982).

For normal grades of hydrated lime, 85% or more of the material passes through a #200 mesh sieve. Finer grades of hydrated lime (99.5% of material passes through a #325 mesh sieve) are also obtainable.

Calcium-Magnesium Hydroxide

The calcium content of dolomitic hydrated lime [$Ca(OH)_2$-Mg) or $Ca(OH)_2$-$Mg(OH)_2$] is lower than that of hydrated lime, having an MgO content greater than 35%. It is manufactured by calcination of dolomite. It has a higher theoretical neutralizing value than an equivalent weight of high-calcium hydrated lime. It generally is less available than high-calcium hydrated lime, and its use in neutralizing acidic surface waters is uncommon.

Sodium Hydroxide

Sodium hydroxide (NaOH) is a common neutralizing agent in industrial processes and in water and wastewater treatment. Common names for NaOH are *caustic soda*, *lye*, and *caustic*. Sodium hydroxide has been used to treat strongly acidic waters in acid mine drainage treatment plants (Lovell, 1973) and ponds affected by acid mine drainage (Olem and Longaker, 1981). Its use in treating mildly acidic surface waters has not been reported. Sodium hydroxide is a powerful neutralizing agent that dissolves instantly in water and can increase pH levels to almost pH 14. Because of its extreme causticity, NaOH is not recommended for the treatment of mildly acidic surface waters due to safety considerations and the possibility of overtreatment.

Silicates

Silicate materials have been used for the neutralization of water. The silicates include basic slag, which is composed of CaO and silicon dioxide (SiO_2) and has the general chemical formula $(CaO)_3$-SiO_2, and olivine [$(MgO)_2$-$SiO)_2$], a naturally occurring silicate material.

Basic Slag

Basic slag is a by-product of the industrial process in which

$CaCO_3$ is used as a slag former to produce metals from ores. Unlike most other industrial by-products sometimes used as neutralizing agents, it is readily available in certain regions.

The dissolution rate of the CaO component of basic slag is a function of the CaO-SiO_2 bond (Grahn and Hultberg, 1975). Because of this bond, basic slag generally does not increase pH to levels above pH 7 or 8. The potential leaching of trace metals may be high, however, for some basic slags, and because the material is a waste product, its use for surface water neutralization may result in resistance from regulatory authorities and the public.

Olivine

Olivine is a naturally occurring silicate mineral that has neutralizing power due to the reaction of acidity with MgO. Its effectiveness as a neutralizing agent increases with increasing magnesium content. It is slightly soluble in water and its dissolution characteristics are similar to those of dolomite. Although olivine is not widely available, its cost is low relative to that of other neutralizing materials. Unfortunately, there appear to be no benefits to using olivine for surface water neutralization; field experiments have generally been unsuccessful (H.U. Sverdrup, personal communication, 1990).

Chemical Reactions Involved in Neutralization

The chemical reactions involved in neutralization of mildly acidic waters by the neutralizing agents described in the preceding section are generally similar. The major chemical reactions for neutralization with calcium carbonate, calcium hydroxide, calcium oxide, sodium carbonate, and sodium hydroxide are described below.

The dissolution of solid $CaCO_3$ in an acidic aqueous solution can be described by three heterogeneous reactions:

$$CaCO_3 + H^+ = Ca^{2+} + HCO_3^- \text{ (pH <6.0)} \quad 2.1$$

$$CaCO_3 + H_2O + CO_2 = Ca^{2+} + 2HCO_3^- \text{ (pH >6.0)} \quad 2.2$$

$$CaCO_3 + H_2O = Ca^{2+} + HCO_3^- + OH^- \text{ (pH >6.0)} \quad 2.3$$

The first reaction (Equation 2.1) dominates at pH values less than about 6.0. Kinetic studies have shown that the reaction rate is transport limited and can be appropriately described in this pH range by applying a film mass transport model in which hydrogen ions diffuse through a boundary layer surrounding the particle.

At pH values above 6.0, the reactions described by Equations 2.2 and 2.3 tend to dominate. Because of the low dissolved CO_2 in acidic lakes or streams, however, the reaction described by Equation 2.2 can essentially be ignored.

Calcium oxide has the same neutralizing effect as $Ca(OH)_2$ (hydrated lime). Calcined lime is hydrated upon contact with water according to the following reaction:

$$CaO + H_2O = Ca(OH)_2 \qquad 2.4$$

Calcium hydroxide has the same neutralizing effect as $CaCO_3$. The species present in solution are determined by the equilibrium with CO_2, as described in the following equations:

$$Ca(OH)_2 = Ca^{2+} + 2OH^- \qquad 2.5$$

$$OH^- + CO_2 = HCO_3^- \qquad 2.6$$

The reaction of sodium carbonate in an acidic aqueous solution is represented by the following reactions:

$$Na_2CO_3 + H_2O = 2Na^+ + HCO_3^- + OH^- \qquad 2.7$$

$$OH^- + CO_2 = HCO_3^- \qquad 2.8$$

$$OH^- + H^+ = H_2O \qquad 2.9$$

The neutralizing effect of sodium hydroxide is the same as that of Na_2CO_3. The species present in solution are determined by the equilibrium with CO_2, as described in Equation 2.10.

$$NaOH + CO_2 = Na^+ + HCO_3^- \qquad 2.10$$

The exposure of the neutralized water to the atmosphere then produces similar reactions to those shown for Na_2CO_3 (Equations 2.7–2.9).

Selection Considerations for Limestone Materials

Results from neutralizing acidic waters over the past 30 years (industrial applications, research, lake liming, and wastewater treatment) indicate that limestone is the material of choice for lake and stream liming. It is easy to handle and noncaustic, obtained relatively free of contaminants, relatively inexpensive, and a natural component of many soils.

The selection of a particular grinding grade of limestone is normally a two-step process (Fraser et al., 1985). First, the method of limestone application is established. Each application method (described later) is generally associated with a range of particle size distributions. Table 2.4 illustrates typical limestone particle size ranges associated with various liming technologies. Particle diameters mentioned herein are surface area-weighted means. The value is weighted according to an individual particle's surface area to account for highly irregular shapes of particles.

Once a liming technique is selected and a range of particle sizes is confirmed, a particular grinding grade is established through economic optimization analyses (Fraser et al., 1985). The general trend in recent years, particularly in the United States, has been toward the use of finer grades of limestone.

Selection Criteria for Lakes, Streams, and Watersheds

Criteria have been established for defining limestone grades that have dissolution and buffering properties appropriate for the liming of lakes, streams, and watersheds.

The following factors are important in the selection of a particular type of limestone for lake treatment: (1) the selected material must be able to produce the desired level of neutralization in the lake, (2) neutralization should be cost efficient, (3) the material should have the potential to dissolve (at least partly) on the lake bottom, and (4) the material should not contain excessive concentrations of nutrients or toxic trace elements (e.g., lead).

The preferred material is generally considered to be limestone with less than 5% $MgCO_3$ content. For lakes of moderate average depth (less than 7–8 m) and acidity (pH 5.4–6.0), the Living Lakes,

Table 2.4 Limestone Particle Size Ranges for Various Neutralization Techniques (Source: Fraser et al., 1985)

	Surface Area-Weighted Diameter[a]	
Application Technique	Acceptable Range	Optimal Range
Lakes		
Aerial application (dry powder)	0.7–18	3.5–7.5
Aerial application (slurry)	0.7–12	3.5–7.5
Boat (dry powder)	0.7–30	7.5–14
Boat (slurry)	0.7–30	7.5–14
Truck	0.7–30	3.5–14
Streams		
Doser (dry powder)	5–13	7.5–12
Doser (slurry)	0.7–5	3.5–5
Diversion well	3–12 mm	6–8 mm
Rotary drum	20–50 mm	38 mm
Limestone barrier	20–50 mm	38 mm

[a] Mean diameter measured in micrometers, unless otherwise indicated. Weighted according to an individual particle's surface area to account for highly irregular shapes of particles.

Inc. (LLI) program operations manual recommends an average particle size in the range of 6–20 μm (LLI, 1987). For deeper lakes (>7–8 m) or strongly acidic water (pH <5.4), limestone with an average particle size in the range of 10–20 μm is recommended. In general, smaller limestone particles are recommended for shallower lakes to ensure dissolution before the particles settle. Limestone materials applied in the LLI program during treatment operations in 1986 are shown in Table 2.5.

The relative distribution of classes of limestone materials used in Sweden for 1983–1987 is shown in Figure 2.5 (Nyberg and Thornelof, 1988). About 80% of the lime used in the Swedish government subsidy program in 1983–1988 was limestone with particle sizes of 0–200 μm and a surface area-weighted average diameter of 14 μm. In 1983, only 60% of the base material applied to lakes in Sweden was in this class; more systems were being treated with larger grades and with materials other than limestone.

For stream neutralization, the following criteria are important in the selection of the specific limestone material:

Table 2.5 Limestone Materials Used by Living Lakes, Inc., During 1986 Operations (Source: LLI, 1987)

Type of Lake Zone	Product[a] (Eco-Cal)	Diameter Range (μm)	Mean Diameter[b] (μm)
Shallow (<10m)	14	0–44	10–12
Deep (>10m)	18	0–70	14–15
	40	0–400	30
Epilimnion	14	0–44	10–12
Hypolimnion	18	0–70	14–15
	40	0–400	30
Sediment	18	0–70	14–15
	40	0–400	30

[a] Commercial product name (Pfizer Chemical Company).
[b] Surface area-weighted mean.

- The selected material should produce the desired level of neutralization rapidly, generally within several minutes.
- The neutralization agent should be cost efficient.
- Excessive residual material in the stream bed is not desired.
- The material should not contain nutrients or toxic trace elements.
- When the material is to be stored as dry powder, it must not be hygroscopic.

Limestone containing less than 5% $MgCO_3$ is generally preferable. For turbulent streams with pH <5.4, material with an average particle size in the range of 3–10 μm is recommended (LLI, 1987). In low-flow streams with pH ≥5.4, the suggested particle size is much smaller: 0.5–5 μm. This same size range is preferred for situations in which dosages exceed 18–20 g/m^3.

For neutralization of watersheds or wetland areas, the criteria important in the selection of the specific limestone material are generally similar to those outlined for lakes. Warfvinge and Sverdrup (1988a) suggested that an extremely fine particle size is not as beneficial in liming forest soils as it is in treating the water column. They recommended a coarse powder with the bulk of the material less than 0.2 mm for soil liming, because larger limestone particles in soil are generally more available for continuous neutralization in watersheds. Larger limestone particles in surface waters, on the other hand, settle and may become unavailable for neutralization.

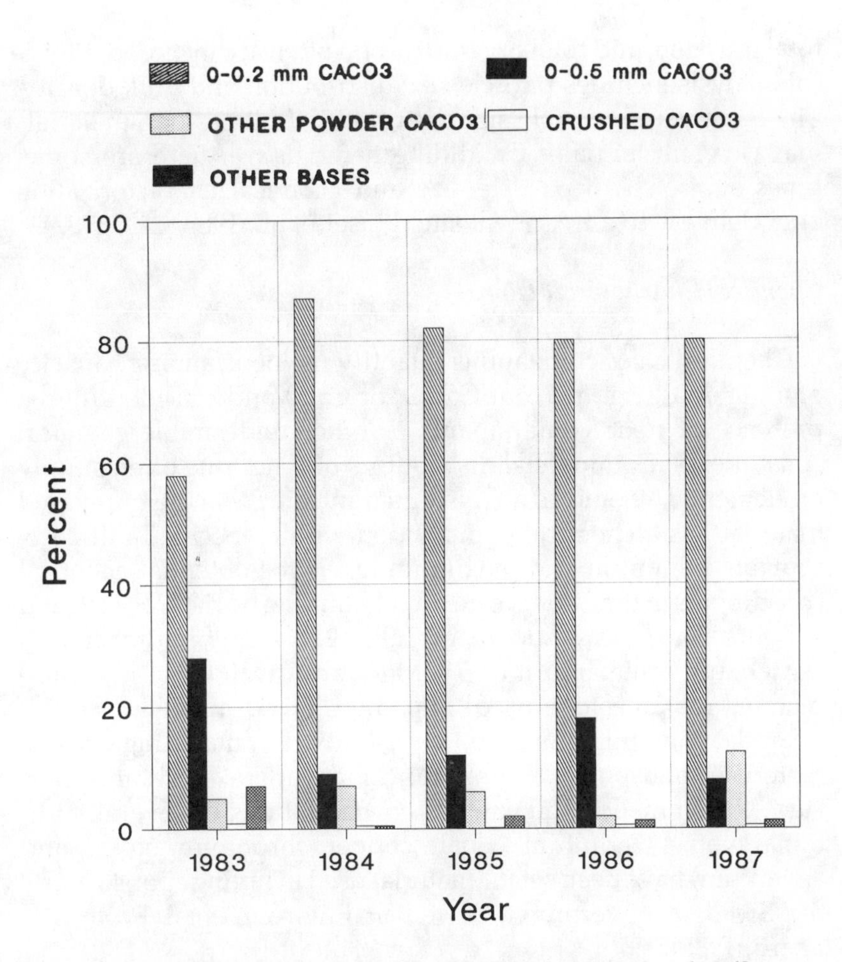

Figure 2.5 Relative distribution of liming materials used in Sweden. (Source: Nyberg and Thornelof, 1988.)

Limestone Quality

Limestone vendors generally adhere to a quality control and quality assurance program consisting of chemical and physical monitoring of the liming material (Pfizer, 1989; Omya, Inc., 1989). Results from vendors' analytical laboratory analyses are often available for review and provide important information to determine whether a material meets selected criteria. Chemical analyses include the variables required by the Food and Drug Administration for food-grade products as outlined in the Food Chemical Codex (LLI, 1987). Additionally, analyses of total phosphorus, total nitrogen, and total organic carbon often are included. Physical analysis includes particle size distribution and bulk density. The quality and particle size distribution of limestone materials may vary substantially. Candidate materials are, therefore, sometimes subjected to particle size and chemical characterization after delivery to a treatment site (Fraser et al., 1985; LLI, 1987).

Chemical Characterization

Chemical characterizations quantify the neutralizing capacity of a material (e.g., percent $CaCO_3$ or CaO) and can identify the presence of toxic contaminants or other undesirable chemical constituents. Commercial distributors of limestone often supply an elemental (chemical) analysis guaranteeing that the purchased material has a chemical composition within specific limits. Acceptable minimum and maximum levels for potential chemical constituents in limestone to be used for liming by the U.S. Fish and Wildlife Service are shown in Table 2.6. A minimum of 70% $CaCO_3$ and a maximum of 5% $MgCO_3$ by weight were recommended for limestone used. Maximum levels also were set for potential contaminants based on fish toxicity, including organic materials, phosphorus, aluminum, manganese, lead, and mercury. Recommended maximum concentrations of several additional metals (cadmium, cobalt, copper, chromium, nickel, and vanadium) have been set for material used in liming sponsored by the Swedish government (Swedish Environmental Protection Board, 1988).

Particle Size

Sverdrup (1984) presented information on particle size distributions derived from commercially available limestone products in Sweden. Figure 2.6 shows distribution curves for 10 limestone products, including limestone slurries and powders and five grades of agricultural limestone. Information on particle size distribution provided by the distributor often includes this type of distribution curve and additional data on bulk density.

The LLI program considered limestone with a surface area-weighted mean particle diameter of less than 14 μm and a maximum absolute particle diameter of 500 μm (curve 6 in Figure 2.6) to be optimal for most lake applications. Usually, only materials

Table 2.6 Acceptable Minimum and Maximum Levels for Chemical Constituents in Limestone Materials Used for Neutralization (Source: Fraser et al., 1985)

Parameter/Element	Minimum Acceptable	Maximum Acceptable	Percent by Weight
$CaCO_3$	X		70
$MgCO_3$		X	5
Organic materials[a]		X	5
Phosphorus		X	0.1
Aluminum[b]		X	1
Manganese		X	1
Lead		X	0.1
Mercury		X	0.0005

[a] Loss at 200°C.

[b] Most often found in the fairly inert form Al_2O_3 .

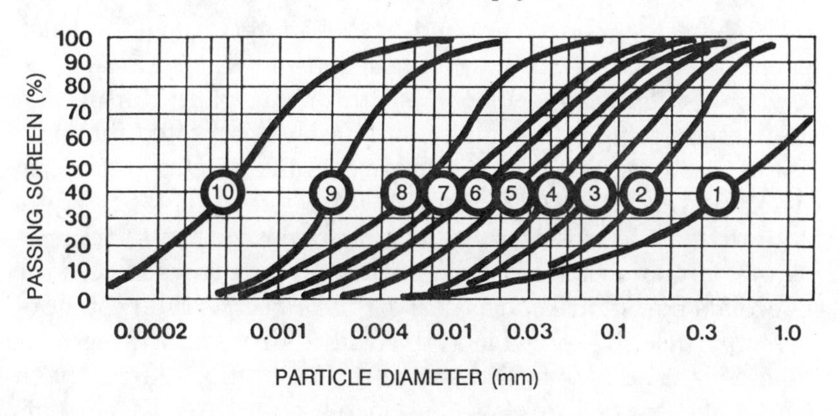

Curve No.	Classification Description
1-2	0-3 mm agricultural limestone
3-4	0-2 mm agricultural limestone
5	0-1 mm agricultural limestone
6	0-0.5 mm powdered limestone
7	0-0.25 mm powdered limestone
8	0-0.05 mm fine powdered limestone
9	0-0.005 mm chalk filler
10	0-0.002 mm marble filler

Figure 2.6 Particle size distribution curves for different commercially available limestone powders used in Sweden. Limestone powders with similar particle size distributions are available in the United States. (Source: Sverdrup, 1984.)

finer than shown by these particle size distribution limits were used in the LLI program for lakes.

Dispersant Materials

Chemical dispersants, such as sodium polyacrylates and guar gum, have been used in helicopter and fixed-wing aircraft applications to keep the limestone material in suspension during transport. Dispersants also allow a more concentrated slurry (thereby reducing transportation costs) and also allow longer storage (Molot et al., 1986).

Dispersants are sometimes used in stream doser slurry storage tanks to keep the particles dispersed before application. Dispersants are usually unnecessary for lake applications in which a slurry box or barge is used. Dispersant chemicals are also sometimes added to the slurry to further reduce the aggregation of limestone particles after doser application to streams (Abrahamsen and Matzow, 1984) and aerial application to lakes (Molot et al., 1984).

The limestone slurry applied by helicopter to Woods Lake, New York was effectively kept in suspension with sodium polyacrylate (Scheffe et al., 1986a). Chemical sampling during and after aerial liming of Bowland Lake, Ontario indicated that only 47% of the limestone dissolved, a lower than expected short-term dissolution efficiency (Molot et al., 1984; Booth et al., 1987). It was assumed that electrostatic forces in the dry powder caused clumping into larger particles that did not dissolve in the water column. A second lake was treated as part of this project with limestone of a finer particle size mixed in a 70% water slurry and with sodium polyacrylate added as a dispersant. Dissolution efficiency in Trout Lake (treated in May 1984) increased to 86%. Dispersants are currently used in the two wet-slurry dosers currently operating in Sweden (E. Thornelof, personal communication, 1990).

Application Techniques

A variety of application techniques have been used to disperse liming materials into aquatic systems. The techniques vary according to the type of liming material used, the type of receptor (whether a surface water or its watershed), and the characteristics of the water body requiring mitigation. The advantages and disadvantages of common techniques used for application of base materials to lakes, watersheds, and streams are summarized in Table 2.7.

The required equipment for application is determined largely by cost, with consideration of the type of system to be treated and its accessibility. For example, the most practical way to lime a remote lake is usually by aircraft, whereas a lake that is accessible by road can be limed by truck, boat, or aircraft. For readily accessible sites, the final selection of equipment options is based on several factors, such as whether the targeted system is a lake or stream, and whether neutralization is to be effected directly by adding liming materials to the water or indirectly by liming the surrounding watershed.

Direct limestone application into a lake or stream is the most commonly used and effective method (Fraser and Britt, 1982). In 1987, direct application accounted for about 85% of the total base material applied in Sweden (Nyberg and Thornelof, 1988). The liming of soils in the watersheds of streams and lakes has received increased attention in recent years, and application techniques are similar to those used for agricultural liming or for liming ice-covered lakes.

Base materials can be applied either in dry or wet (slurried) form. Sverdrup and Warfvinge (1988a), comparing lake liming techniques, concluded that adding limestone as dry powder was much less efficient than adding it in a wet slurry. This is the result expected because short-term dissolution efficiency is higher for slurried material.

Recommended limestone particle sizes differ according to the method of application (Fraser et al., 1985, Table 2.2). Additional information on base materials used for each technique and a description of the technology, selected case studies, and design criteria for each technique are given in the following subsections.

Techniques Used in Lake Applications

There are four common techniques for applying base materials to lakes: by boat or barge, truck or tractor, helicopter, or fixed-wing aircraft.

Boat and Barge

Boat applications include any method in which a boat or barge is used to spread liming materials over the surface of the lake or to inject the materials into the sediments. The efficiency of this

Table 2.7 Advantages and Disadvantages of Liming Techniques

Receptor	Technique	Relative Use	Advantages	Disadvantages
Lake	Boat	Common	Simple, effective, and accurate. Allows different distribution options. Commercial applicators exist.	Not very practical for remote sites.
	Truck or tractor	Occasional	Can distribute materials more quickly than boat method. Simple method for treating ice-covered lakes.	Not practical where road access is limited. Able to treat only small lake from shore or ice-covered lake. Ice application may be dangerous.
	Aircraft	Occasional	Allows use for remote sites. Less labor intensive once loaded. Distribution of materials is quicker than by boat or truck.	Expensive.
Stream	Doser	Common	Allows more precise maintenance of pH than other methods. Can be computer controlled.	Regular maintenance required. Relatively expense to construct and maintain. Sophisticated systems require electricity.

	Diversion well	Occasional	Simple and relatively inexpensive. Can be regulated by flow.	Not possible in slow flowing streams. Requires considerable piping hook-up.
	Limestone barrier	Uncommon	Simple and relatively inexpensive. Limestone is present long after application.	Technique fails during high flow periods. Must have easily accessible site. Limestone usually becomes coated with sediment, biological growth, and precipitates, and rendered ineffective.
	Rotary drum	Occasional	Allows effective maintenance of pH. Continual abrasion of material prevents coating by sediment and precipitates.	Regular maintenance required. Not possible in slow flowing streams. High construction costs.
Watershed	Truck or tractor	Common	Simple, fast, often less expensive than aircraft.	Not practical for remote or steep sites; road access required.
	Aircraft	Common	Simple, fast, allows use at remote sites. Can deliver material where needed independent of roads.	Expensive.

method is generally limited by the carrying capacity of the equipment used. Sometimes a second boat transports neutralizing materials from the shoreline to replenish the supply. A variation of this method involves the continuous supply of material to the boat from shore through a pressurized hose.

The application of liming materials by boat (or barge), which is probably the least expensive method for accessible lakes, is now the most commonly used method. About 55% of the total neutralizing material applied in 1987 for the Swedish operational liming program was from pontoon boats (Figure 2.7).

Emptying bags of powdered limestone by hand from a slowly moving boat is the simplest form of boat treatment. It is the least effective of the boat application methods. More complex technology has been developed and applied to mitigation projects that allows more controlled addition of base material (Lepley, personal communication, 1989). Descriptions of several boat- or barge-based application methods are presented in the following sections, including commercial systems available in Sweden (IMEK AB) and the United States (Sweetwater Technology Corporation).

Manual Dry Spreading. This approach involves the manual distribution of limestone from a boat, barge, or platform towed behind a boat. The material is often distributed from the bow of a moving vessel so that turbulence from the propeller increases mixing and distribution (Fraser et al., 1985). Although this technique is one of the simplest methods of liming a lake and one that requires little equipment, it is the most labor intensive and time consuming of the boat distribution methods and results in the least effective control of dosage and dissolution of neutralizing material. Because of the physical labor involved and boat carrying capacities, the maximum application rate from a single boat is about 1 tonne/hr. This method is not used in any operational liming program.

Several ponds in Massachusetts were treated by manual spreading of dry agricultural limestone (Bergin, 1984; Fraser et al., 1985). Asnacomet Pond, Hubbardston, Massachusetts was treated in October 1982 by using this technique. Agricultural limestone with 40% of the material less than an average particle diameter of 28 μm was selected at a dosage of 10 g/m^3. Most of the limestone was spread in the littoral zone, where the depth was generally less than 7 m. It was distributed manually from bags over the side of an outboard motorboat (5 m long) and was mixed in the wake of

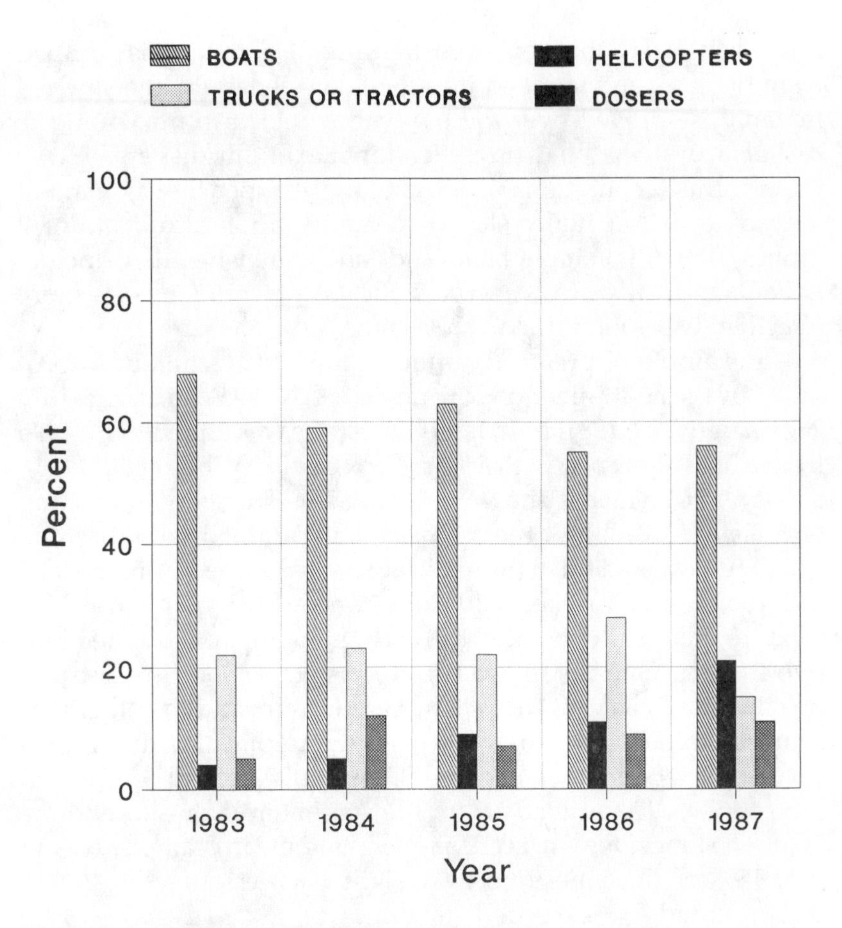

Figure 2.7 Relative distribution of different liming application methods in Sweden expressed as percentages of the total quantity of material applied, 1983–1987. (Source: Nyberg and Thornelof, 1988.)

the vessel. The 51.5-ha pond required a calculated dosage of 45.4 tonnes of limestone. The mean pH and alkalinity were 5.15 and 120 μeq/L before treatment and 6.54 and 120 μeq/L after treatment (as determined from five samples collected in 1983). No reason was provided for the observed similarity in pretreatment and posttreatment alkalinity. It is likely that the pretreatment alkalinity reported was too high relative to the pH.

Slurry Box Spreading. This technique is similar to the manual dry spreading method but includes equipment for the slurrying of dry-powdered material in a tank filled with lake water and carried aboard the boat or barge (Figure 2.8). The slurry is pumped from the tank and sprayed into the lake. Generally, the slurried material is sprayed over the bow, allowing it to mix from the turbulence of the boat propeller for maximum dispersion. Although this technique is as simple and inexpensive as manual spreading, it is equally slow (maximum application rate of 2 tonnes/hr with a single boat) and labor intensive. Its principal advantage relative to the dry application technique is its more effective dissolution due to slurrying.

An example of practical application of this technique is provided by the neutralization of Sandy Lake, Nova Scotia. From July 28 to August 28, 1981, powdered limestone was added to the lake from a boat equipped with a slurry box device (White et al., 1984), a diagram of which is shown in Figure 2.9. The fiberglass mixing tank held 3000 L, and the material was pumped with an 8-hp, gasoline-powered trash pump. Water was pumped from the lake into the tank and mixed with limestone added from bags. This mixture then was circulated through the pump and divided into two streams. One stream was pumped with a short fire hose and nozzle into the lake at the rate of 1136 L/min, and the other was returned to the tank. An overflow, fitted to the tank, discharged excess slurry over the stem of the 5.2-m boat. A second boat ferried limestone in 900-kg loads from the staging area on shore to the boat. The limestone surface area-weighted mean diameter was 18 μm; 15% of the limestone by weight was less than 10 μm in diameter and 15% exceeded 100 μm.

Over the 1-month period, 135 tonnes of limestone was spread evenly over the surface of the 74-ha lake.

Barge and Pressure Tank. This technique, in concept, is similar to the slurry box technique, but is much more automated and requires less labor. For this technique, pontoon boats or barges are equipped with large, pressurized storage tanks, compressed air blower systems, slurrying units, generators, and pumps to supply lake water to the units. Pressurized tank trucks, located at the docking area, pneumatically pump dry-powdered limestone into the storage tank on the boat or barge. Lake water is pumped into a small slurry tank and "flash-mixed" with limestone from the pressurized tank. The slurry mixture is then sprayed through fire hoses onto the lake surface. The method has an estimated application rate of 8–10 tonnes/hr from one boat or barge.

Figure 2.8 Spreading of limestone from a slurry box on a boat. (Photograph courtesy of Minnesota Department of Natural Resources and U.S. Fish and Wildlife Service.)

This technique has been commercially available in the United States and Sweden since about 1983. Lake Masen, in the province of Halland, Sweden, was treated in November 1984 by this technique (Fraser et al., 1985). The 420-ha lake was treated with 1700 tonnes of limestone with a surface area-weighted mean particle diameter of 12 μm. The short-term dissolution efficiency was 63%. The pH and alkalinity of Lake Masen were 5.3 and -100 μeq/L before treatment and 6.8 and 100 μeq/L after treatment.

The only commercial barge/pressure tank treatment device currently operating in the United States was designed and built by Sweetwater Technology Corporation (R.H. Lepley, personal communication, 1989). The vessel (a pontoon boat) is 8.5 m long,

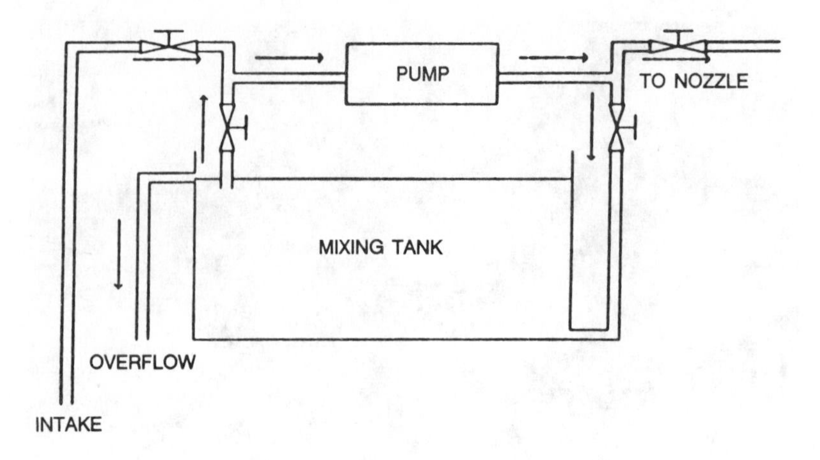

Figure 2.9 Diagram of slurry box spreader. (Source: White et al., 1984.)

2.4 m wide, and 1.2 m deep and powered by a 100-hp outboard motor. Its steel tanks have a holding capacity of about 1.1 tonnes. Slurried material is applied to the lake surface through retractable spray arms that cover a width of 18 m (Figure 2.10). The Sweetwater barge uses a patented process for delivering slurried materials to the surface of the water body. Distribution is controlled by an onboard computer that calculates applications on a volumetric or surface-area basis. For volumetric application, both speed and water depth are read directly by the computer, which then calculates the proper rate of chemical distribution. For distribution on a surface area basis, only barge speed is read by the computer. Flowmeter readings are monitored by the computer, which then regulates the distribution valves until the proper slurry flow is obtained.

The barge is also equipped with an onboard navigation system that permits radio signals to be received on land and retransmitted to the barge computer to determine its location. The barge operator then steers the barge to avoid overlap or skipping of treatment paths.

The device was demonstrated in fall 1987 at two sites (Brocksen and Emler, 1988; Adams and Brocksen, 1988). White Deer Lake is a shallow (maximum depth 4.7 m) 19.8-ha lake in Pike County, Pennsylvania. Because of the shallow depth, the lake was treated with a very fine limestone powder (surface area-weighted mean diameter of 4 μm), as well as with 18-μm diameter limestone.

Figure 2.10 Commercial barge pressure tank application of limestone in the United States. (Photograph courtesy of Sweetwater Technology Corporation.

Using the barge and pneumatic application system, the crew applied 16.1 tonnes of material within 5 hr. Wallum Lake is a large (130 ha) recreational lake on the border between Rhode Island and Massachusetts. Staging areas were set up on two sides of the lake, one area in each state. A total of 99.5 tonnes of limestone was applied within 16 hr.

A modification of the barge and pressure tank system was developed in Sweden, in which the boat is outfitted with a long rubber hose and the neutralizing material is continuously pumped as a slurry from shore to the boat (Sverdrup, 1984). The process, known as the Kalcade method, has been successfully operated at distances of several kilometers. Sverdrup (1984) reported that the method may be a viable alternative to aerial application for lakes without direct road access. The peak delivery of the system was reported to be 25–40 tonnes/hr. A somewhat more dilute slurry must be used than that used in the other barge and pressure tank techniques because of the long pumping distances. This process was applied to the treatment of Lake Stora Lee, a large (14,300 ha) lake in Varmland, Sweden. In 1982, a total of 9000 tonnes of limestone was applied from a staging area located 1.8 km from the boat. The method is no longer commercially available, apparently due to difficulties encountered handling the long hose (E. Thornelof, personal communication, 1990).

Sediment Injection. This technique consists of injection of base materials into the sediments of acidic lakes. The process has been tested experimentally on at least two occasions, but has not been used operationally.

Neutralizing materials used in the process include limestone, hydrated lime, and soda ash (Fraser et al., 1985). The treatment has been reported to increase sediment base ion exchange with the acidic water column (Lindmark, 1982, 1985). Calcium or sodium ions in the sediment are released in exchange for hydrogen ions from the water column. This results in a gradual change in pH and alkalinity of the lake water and an increase in base cation and ANC supply to the water column during spring and fall lake turnover. The technique has also been shown to release phosphorus from the sediments to the water column, resulting in increased productivity and subsequent benefit to the fisheries.

Fraser et al. (1985) reported that this technique is generally limited to small, shallow lakes and ponds with soft organic sediments and adequate road access for transport of materials and application equipment. It may also be a useful method for treating lakes with high flushing rates (Lindmark, 1982). In laboratory experiments, this treatment has been shown to be about five to seven times as efficient as adding limestone on an equivalent basis (Ripl, 1980).

The Contracid™ technique was developed at the Institute of Limnology (Lund University, Sweden) in cooperation with a commercial vendor, Atlas Copco Corporation (Figure 2.11). A sediment harrow—a structure similar to a regular farming harrow—is dragged along the bottom of a lake, raking the top 20 cm of the sediment and injecting Na_2CO_3 with a compressed air dispersal system (Stoclet, 1979).

The effectiveness of the Contracid™ method was tested at Lake Lilla Galtsjon, a small (14.7 ha), shallow acidic lake in a readily accessible part of southeastern Sweden (Lindmark, 1982, 1985). Of the total lake sediment, 25% was treated in October and November 1980 with 270 g Na_2CO_3/m^2 of lake surface area. A total of 40 tonnes Na_2CO_3 were applied in a 10% lakewater solution. This treatment increased the sediment's base ion exchange, resulting in an increase in the lake pH from 5.00 to 6.66 and alkalinity from 0 to 120 μeq/L (Lindmark 1982, 1985). Fraser et al. (1975) reported the use of less sophisticated systems of similar function in privately owned ponds in the Adirondack region of New York;

Figure 2.11 Contracid™ method of injecting base material into the sediment of acidic lakes. (Drawing by Atlas Copco Corporation.)

however, the technique has the potential to disrupt the benthic community and to increase water column turbidity.

Truck or Tractor

Land vehicles such as trucks and tractors have been used to transport and distribute liming materials to lakes; a manure spreader or similar device is used to spread the material on ice or by using compressed air to blow the material onto the target areas from shore. Application to ice may be an appropriate method in northern climates. Blowing dry material is not a viable method for lakes because of poor control of dosage and dissolution of neutralizing material.

One system was extensively used in Sweden in the late 1970s and early 1980s by adapting equipment originally designed for cement construction projects (Fraser and Britt, 1982). A large trailer (about 40-tonne capacity) near the lake serves as a staging area. Limestone is spread along accessible waterways, including drainage ditches, by using a truck with a pair of 6-tonne tanks and a blower system. Two men can apply about 80 tonnes in 10–12 hr. Dry-powdered limestone is spread very inefficiently with this method and it is not recommended.

Liming materials have been applied to lake ice with tractors at several sites, particularly the northern regions of Sweden and Norway. Tractors are generally used to haul pallet loads of bagged limestone from the nearest road to the lake. A manure spreader is sometimes used to apply the limestone to the lake ice (Figure 2.12). Application of limestone to lake ice is a viable method that is used in the Swedish and Norwegian operational liming programs.

Helicopter

The application of liming materials by aircraft is usually the most efficient and cost-effective method for remote waters. Helicopters can be used to treat remote or inaccessible lakes by spraying dry-powdered or slurried limestone, slowly traversing the lake target zones (Figure 2.13). Commercial systems for application of base materials by helicopter are available in Sweden.

In the helicopter method for applying dry-powdered limestone, more than one bucket is generally used. Buckets are alternately filled with dry limestone powder at a staging area, and a filled bucket is transported to the target water (Figure 2.14). The helicopter traverses the lake at an altitude of 50–200 m, and a door in the bottom of the bucket is opened electronically, releasing the load. This technique may inaccurately distribute material under windy conditions, releasing fine particles that may obscure the pilot's vision, and possibly damage the helicopter rotor blades and shaft. Prudent pilots carefully consider wind speed and direction before liming.

In the slurry technique, a helicopter is fitted with a storage tank, which is filled with slurried limestone from a tank truck, and a spray nozzle (similar to that used for hydroseeding of clear-cut or strip-mined areas). The slurry can be sprayed more accurately than the dry powder, because the helicopter can hover or fly slowly over the site. Fraser et al. (1985) reported that a helicopter with a 1-tonne storage tank can apply an estimated 5–7 tonnes/hr

Figure 2.12 Truck and spreader for distributing limestone on the ice-covered surface of a lake in Massachusetts. (Photograph courtesy of Massachusetts Division of Fisheries and Wildlife.)

if the target lake is within 10 km of the staging area (LLI, 1987).

A chemical dispersant is often mixed with the limestone as the slurry is prepared to keep the material in suspension in the storage tank. Sodium polyacrylate has often been used as the dispersant material for aerial applications of limestone slurry (Fraser et al., 1985; Booth et al., 1987).

Blake (1981) summarized the experience of the New York State Department of Environmental Conservation in using helicopters to apply lime, primarily to remote ponds in the Adirondack Mountains. Trials with bagged agricultural limestone opened in the helicopter and the material slowly released were reported as unsuccessful, as was the use of hydrated lime mixed as a slurry and dispersed from a fire-fighting bucket. A technique for dry dispersal of material was, however, reportedly successful: a bucket of about 1-tonne capacity was first loaded on the ground with powdered agricultural limestone. This delivery system also used a ground crew to load one bucket at a staging area while a second bucket was emptied over the target pond.

Fraser et al. (1985) reported on the use of trucks in Sweden that were specially designed to hold three 500-kg capacity buckets and

Figure 2.13 Helicopter application of limestone in the United States. (Photograph courtesy of Living Lakes, Inc.)

to allow the helicopter to land on a platform constructed on top of the truck cab. After landing on the platform, the helicopter is secured to one of the buckets, which was pneumatically filled with powdered limestone from a nearby tank truck. With this type of system, a single helicopter can apply material at the rate of 5–7 tonnes/hr.

A helicopter was also used for a project conducted by the

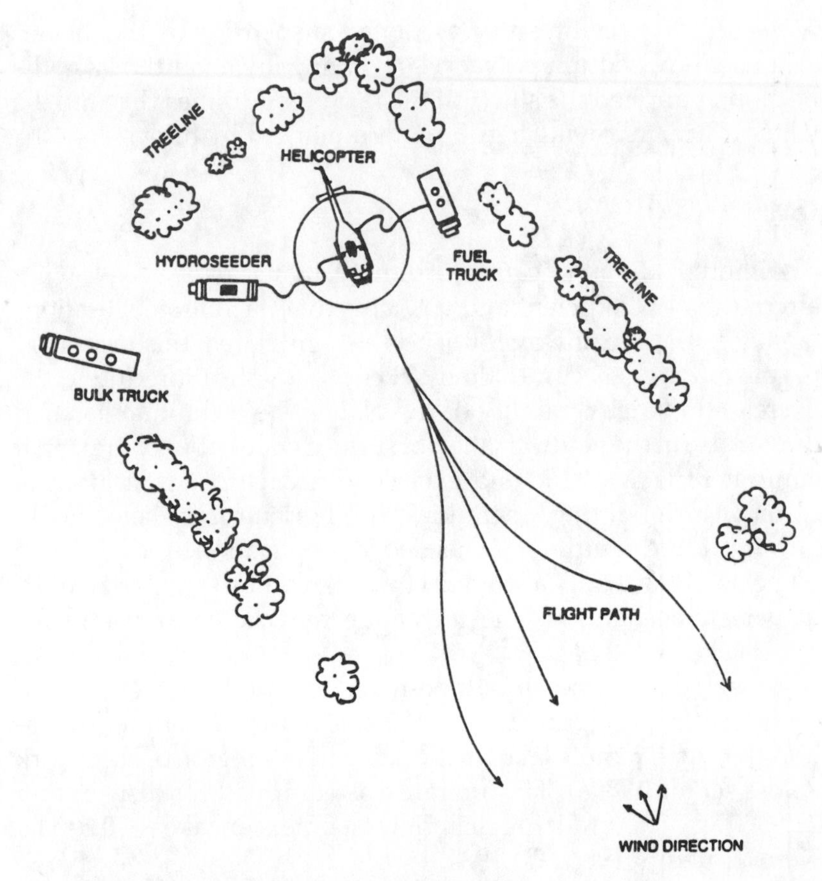

Figure 2.14 Typical helicopter staging area. (Source: Living Lakes, Inc., 1987.)

Swedish Water and Air Pollution Research Laboratory (Fraser and Britt, 1982). During September 1979, the helicopter applied 30 tonnes of powdered agricultural limestone within 4 hr in a strip 10 m wide and 600 m long, following the course of a small stream leading into a lake.

The first reported application of limestone slurry by helicopter in the United States was the neutralization of Woods Lake, New York on May 30–31, 1985 (Scheffe et al., 1986a). The helicopter was equipped with a 1.1-tonne storage tank and a revolving multiple-nozzle sprayer attached to the bottom of the helicopter. The storage tank was filled from a tank truck in a staging area about 6.4 km from the lake. A total of 22.7 tonnes (dry weight) of slurried limestone powder was distributed. The material had a surface area-weighted mean particle diameter of 19 μm, was composed of

71% solids by weight, and was kept in suspension by the dispersant chemical sodium polyacrylate. Flying low over the lake, the helicopter applied the slurry at the rate of 5–7 tonnes/hr onto the lake surface, following a grid pattern marked by buoys.

Fixed-Wing Aircraft

Although the application of liming materials by fixed-wing aircraft may be efficient and cost effective for remote situations having a nearby runway for use as a staging area, this technique is generally less accurate than helicopter application due to the decreased maneuverability of the plane. The technique has had limited use in the neutralization of surface waters, but commercial equipment is available and experience exists for application of water in fire-fighting operations and agricultural chemicals in farming. Little equipment is needed for application by aircraft: the plane must have only a storage bay or tank for the material and a dispersal mechanism. The technique may be most useful for larger lakes (> 50 ha).

A small fixed-wing aircraft designed for application of agricultural fertilizers (Figure 2.15) was used in private liming operations for small remote lakes in the Adirondack region of New York (Gloss set al., 1989a). The aircraft was equipped for distributing 225-500 kg of neutralizing material at a time. Ten lakes of 0.5–5 ha were treated in 1983 and 1984.

Powdered limestone was applied aerially to two lakes in Ontario, Canada (Booth et al., 1987; Molot et al., 1984). In August 1983, 83 tonnes of dry-powdered limestone were applied to Bowland Lake (a remote lake with no road access, 70 km north of Sudbury) from a Canso bomber (Figure 2.16) The Cansco is a twin-engine amphibious plane used to dispense water in forest fire control operations. For this operation, the plane held about 2.4 tonnes of dry material, but had a maximum capacity of 3.8 tonnes. The amount of dry material was limited by the volume of the limestone; the full capacity was possible when a 70% slurry was used. Later use of slurry allowed more limestone to be carried.

Roundtrip flight time between the lake and the Sudbury airport was 40 minutes, and the loading operation took approximately 25 min. The material was dropped from heights of 15–25 m at flying speeds averaging 145 km/hr. Drops were confined to the middle of the lake to avoid shoreline hazards, and deeper waters received more drops than shallow water.

Figure 2.15 Application of limestone by small fixed–wing aircraft designed for applying agricultural fertilizer.

Figure 2.16 Application of limestone by large fixed-wing aircraft designed for fire-fighting operations. (Photograph courtesy of BAR Environmental.)

Techniques Used in Stream Applications

The design and operation of stream liming devices have relied in part on experience gained with systems used for neutralization of acid mine drainage, industrial water and wastewater treatment, and related industrial operations (Lovell, 1973; Penn Environmental Consultants, 1983; Wilmoth, 1977). The technology has advanced rapidly over the last few years; commercial vendors now market stream liming devices in Norway (Norcem A/S), Sweden (Boliden AB; Boxholm AB; Cementa Movab AB; Enerchem AB), and the United States (Sweetwater Technology Corporation). The estimated numbers of each type of commercial stream liming device currently operating in the United States, Canada, and Scandinavian countries is shown in Figure 2.17.

Four common techniques for application of base materials are described for flowing waters: doser, diversion well, limestone barrier, and rotary drum. Limited applications of base materials to flowing waters by truck or tractor and aircraft have been conducted as previously described. Conditions appropriate for selection of a particular stream treatment device are summarized in Table 2.8.

Doser

Dosers are considered here to be fully automated mechanical devices that release powdered or slurried neutralizing material into streams. Other similar devices that use limestone aggregate, such as rotary drums an diversion wells, are considered separately.

State-of-the-art dosing devices are designed to dispense dry powder or slurried limestone directly into the target streams. The distribution of neutralizing agents from dosers powered by electricity or battery is controlled automatically by microprocessors programmed to calculate appropriate dosing rates from remotely monitored water quality or hydrological variables. Dosers powered by water flow distribute neutralizing material at rates that vary with the flow. Four major categories of dosers are used—electrically powered dry-powder dosers, electrically powered wet-slurry dosers, battery-powered dry-powder dosers, and water-powered dry-powder dosers.

Electrically Powered Dry-Powder Dosers. Electrically powered dry-powder dosers consist of a silo containing a limestone storage bin and an electrically powered feeder screw, an automated dose

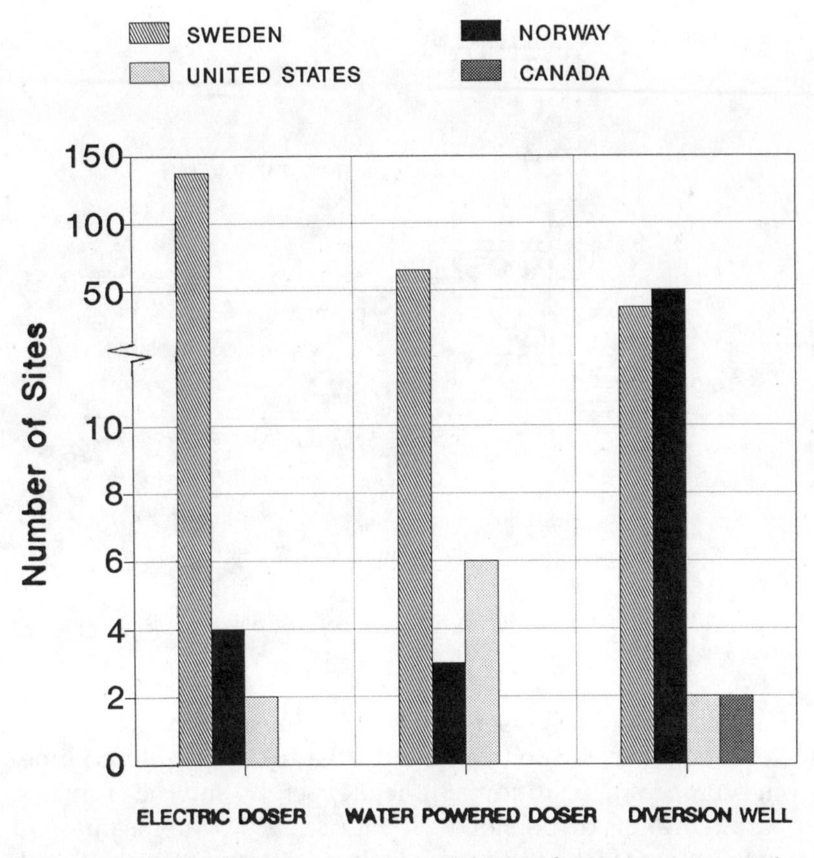

Figure 2.17 Estimated number of commercial stream liming systems operating in the United States, Canada, Norway, and Sweden in 1990.

control mechanism, and a distribution pipe that dispenses the dry powder into the stream (Figure 2.18). Typically, the limestone feed is automatically regulated by the water level, a surrogate for water flow. As the water level fluctuates, various amounts of dry-powdered limestone are transported from the storage bin by the feeder screw to a conveyor belt. This belt, located within a pipe, transfers the neutralizing material from the silo to the point of distribution above the center of the stream.

Electrically powered dry-powder dosers may produce the same slurry output produced by the wet slurry dosers described in the next section. The *dry* and *wet* distinctions refer to the material storage characteristics and not to the output. Some dry-powder dosers slurry the material with stream water on site

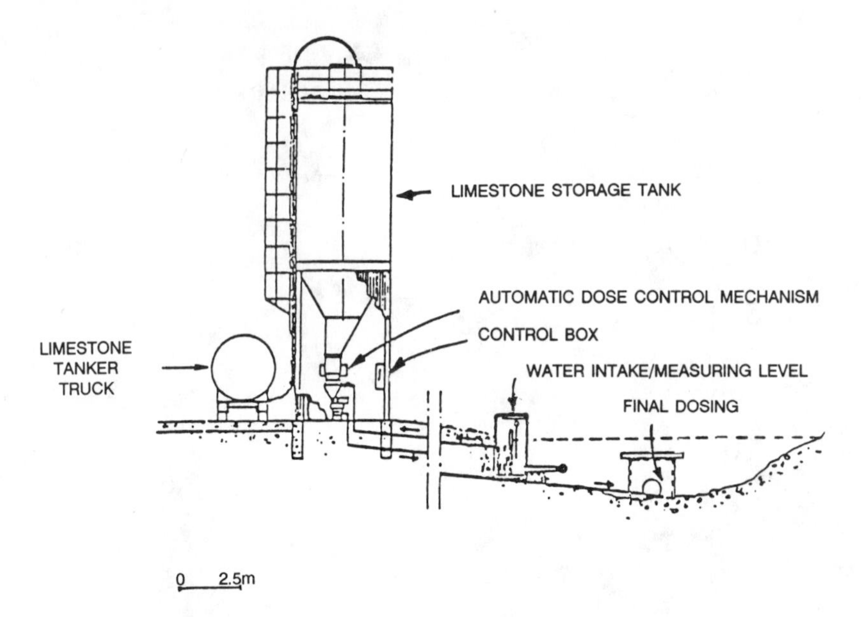

Figure 2.18 Diagram of an electrically powered dry-powder doser. (Source: Ostensson, 1983.)

before dispensing it (Fraser et al., 1985). Such dosers consist of a water intake well, a silo (containing a large dry-powdered limestone storage bin, an automated feeder screw, and the slurrying apparatus), and a dispersion well (Figure 2.19). Water is pumped from the stream to the slurry tank in the silo, where it is mixed with the limestone powder. The resultant slurry is pumped to the dispersion well in the stream bed, mixed with additional stream water, and dispersed just below the stream surface. A typical commercially available system is shown in Figure 2.20.

Electrically powered dry-powder dosers are erected on a concrete foundation near the stream to be treated. The lower section of the storage bin or silo is usually tapered to prevent the build up of powder on its walls. Additionally, the silo is equipped with a hammer, or similar device, that periodically strikes (or vibrates) the wall to loosen any compacted material. Fraser et al. (1985) described the following design requirements and limitations for a typical dry-powder doser.

> The silo is fabricated from metal or other similarly strong and weather-resistant materials. These materials can be modified from commercially available insulated storage bins used in

Table 2.8 Comparison of Liming Techniques for Acidic Streams (Source: Fraser et al., 1985)

Factor	Electrical Dosers	Battery-Powered Doser	Water-Powered Doser	Diversion Well	Rotary Drum	Limestone Barrier
Maximum flow (m^3/sec)	25	8	3	1	1	Varies
Stream head required (m)	0	0	1.0	>1.3	>1.5	0
Moving parts	Yes	Yes	Yes	No	Yes	No
Electrical power required	120/240 V AC	12 V DC	None	None	None	None
Type of neutralizing agent normally used	Slurried limestone powder	Dry limestone powder	Dry limestone powder	Crushed limestone	Limestone aggregate	Limestone aggregate
Other limitations	Sensitive to thunderstorms and power failures. Wet-slurry intake piping may become clogged.	Sensitive to freezing (below -5°C).	Sensitive to freezing (below -5°C).	May not perform well during low or high flows or for highly acidic systems. Limited ability to increase pH more than one unit.	Sensitive to freezing (below -5°C). Generally higher initial cost.	Limestone usually becomes clogged and coated with sediment, biological growth, and precipitates.

Figure 2.19 Electrically powered dry-powder doser. (Photograph courtesy of Per Ostensson.)

municipal water treatment plants or in the storage of agricultural products. In northern climates, all pipes to and from the doser are insulated or heated to prevent ice formation during extended subfreezing temperatures.

The microprocessor and other electrical components are also typically enclosed within some type of insulated structure connected to the silo to protect against inclement weather.

Any finely ground limestone powder can be used in the dry-powder doser. Powders having a surface area-weighted average particle diameter less than 12 μm, however, can collect on the sides of the storage bin during periods of high humidity (Fraser et al., 1985).

Large electrically powered dosers can be used to treat streams with a wide range of flow rates. They have been used successfully on systems with flows as high as 25 m^3/sec, as well as on smaller streams with mean annual flow rates of less than 5 m^3/sec.

The minimum required storage capacity is typically based on the amount of material needed for 30 hr of continuous dosing at

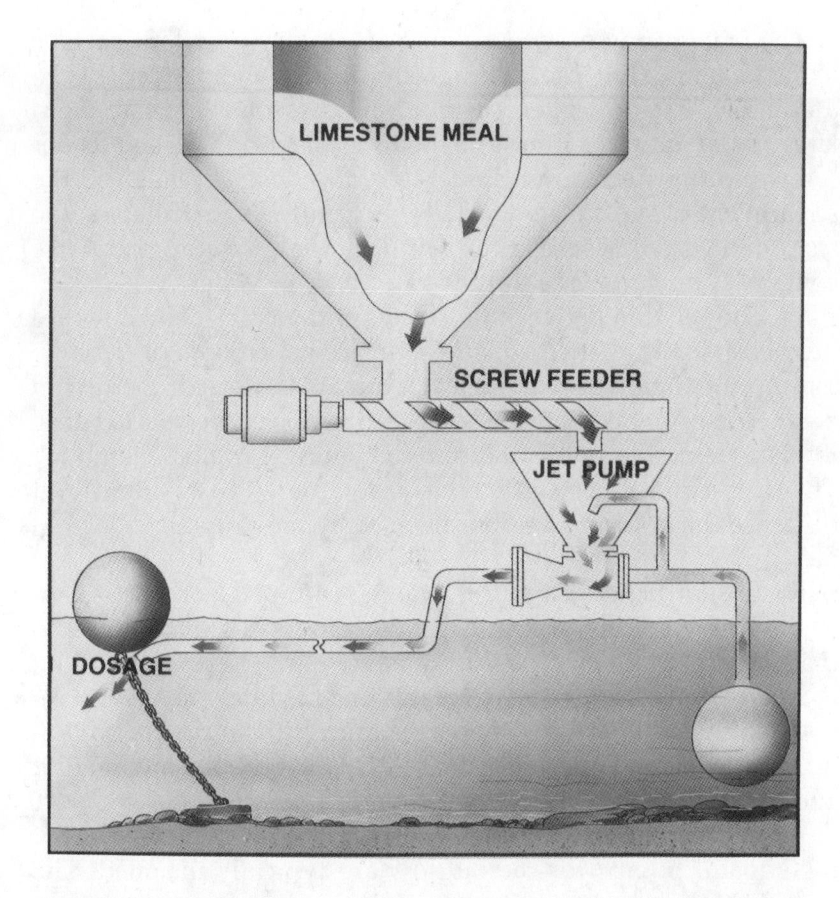

Figure 2.20 Diagram of an electrically powered dry-powder doser dispensing limestone as a wet slurry. (Source: Cementa Movab, 1989.)

the maximum rate of flow, allowing 6 hr for the storage bin to be filled daily by a local distributor, if needed. For existing Scandinavian sites, the volumes of the storage bins range from 4 to 80 m^3 (5- to 90-tonne capacity). The optimum storage capacity can be determined from material and transport cost functions and a knowledge of the total annual base demand of the stream. The amount of limestone in the storage bin is monitored by acoustic recording gauges or other similar devices. Limestone is ordered whenever the quantity of neutralizing agent is reduced to a predetermined threshold. Limestone powder is removed from the bottom of the silo system by feeder transport screws. These screws are typically installed in tandem so that one remains available for use if the other becomes clogged or damaged by foreign objects in the limestone.

The maximum dosage rate for existing dry-powder dosers is about 1 tonne/hr. The appropriate dose is maintained by an automatic control mechanism (microprocessor), which also regulates the quantity of limestone being transported by the feeder screws to the slurry unit. Preferably, the dose is related to the stream water level, from which the streamflow is calculated. The stage of the stream is the critical input variable for automatic dose control. A stream stage-to-flow-rate relation is derived empirically. The relation between flow rates and stream pH values are derived for different seasons from historical records or baseline monitoring data. The desired water quality targets for the stream are then used to calculate the doses. A titration curve relating dose to pH is derived empirically from a chemical equilibrium model. The dose control system is then programmed to calculate and dispense the appropriate dose from a stream-gauging input signal.

For dosers dispensing slurry, water is pumped from the stream to the slurry mixing unit. A constant flow rate is maintained so that the dry content of the slurry never exceeds 10%. Slurrying creates turbulence and shear stresses that facilitate the separation of agglomerated particles, thereby significantly improving dissolution efficiency. Turbulence is generated in both the mixing unit and in the discharge pipe to the stream.

If the doser is situated only slightly above the surface water of the stream, pumps for the outflow are typically installed. The specific location of the outlet dispersion well in the stream is not crucial for the performance of the doser but should be situated far enough below the surface of the stream to prevent damage from freezing. As a rule, the outlet well on the bottom of the stream should have its opening directed toward the surface of the stream to maximize the mixing of the limestone with the stream water and to minimize the settling of undissolved particles.

The storage bin is pneumatically filled at the top from a bulk-pressurized tank truck (such trucks typically hold 30 tonnes of material and are commonly used for transporting materials in the cement, pulp and paper, and milk industries). Typically, the storage bin is refilled weekly, except during high flow periods (e.g., spring), when refilling may be required at 1- to 2-day intervals.

Operation of an electrical powered dry-powder slurrying doser was begun on the headwaters of the Fyllean River near Ryaberg,

Sweden in September 1982 (Fraser et al., 1985). The doser was designed to neutralize a downstream chain of lakes that provide nursery areas for sea-run brown trout. This doser, in conjunction with a wet-slurry doser at nearby Marback, Sweden, also helped to reestablish the Atlantic salmon farther downstream. The installation uses limestone powder with an average particle diameter of 12 μm, and the doser dispenses 8–21 g/m^3 for flows ranging from 1.5 to 15 m^3/sec. Dosage requirements for different flow rates were determined by analyzing and titrating river water collected upstream and downstream from the silo. Limestone requirements were about 1300 tonnes/year for the first and second years of operation.

Electrically Powered Wet-Slurry Doser. A wet-slurry doser is an automated device similar in operation to a dry-powder slurrying doser. The wet-slurry doser system stores and dispenses a commercially prepared slurry of extremely fine limestone powder into streams, whereas the dry-powder slurrying doser stores dry-powdered limestone that is slurried just before it is dispensed. The material from a wet-slurry doser is pumped directly from a slurry storage tank to a Y connection, where it is diluted with water before dispersal. An automatic dosing unit regulates the amount of slurry released through the pipe into the stream. As in the dry-powder dosers, the dose is based on flow rate and calculated by a microprocessor.

Fraser et al. (1985) described design requirements and limitations for a wet-slurry doser that stores a commercially prepared slurry. A diagram of a typical wet-slurry doser system currently used in Sweden is shown in Figure 2.21. Designs that are unique to these dosers, compared to dry-powder dosers, are discussed below.

The preferred neutralizing material is a commercial limestone slurry, with a 70% dry content and an average particle diameter of less than 3 μm. Sometimes a dispersant such as sodium polyacrylate is added to reduce the settling rate and to facilitate the dissolution of the particles.

The typical capacity of existing Swedish storage tanks ranges from 45 to 80 m^3. Acoustic gauges or piezoelectric pressure sensors can be used to monitor the slurry level in the tank. The

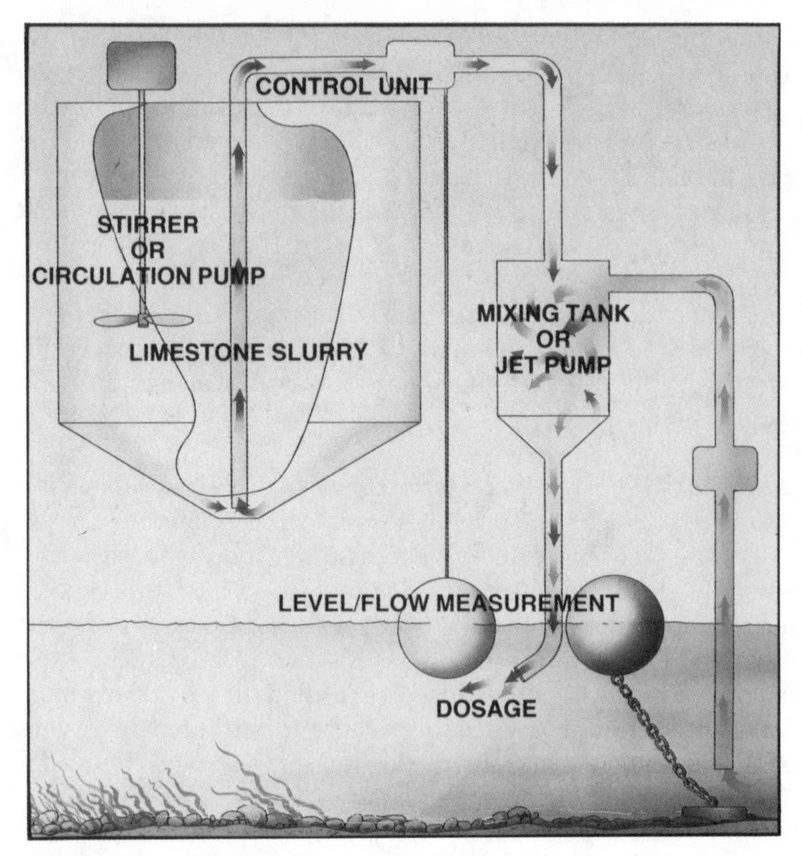

Figure 2.21 Diagram of an electrically powered wet-slurry doser storing a commercially prepared limestone slurry. (Source: Cementa Movab, 1989.)

slurry in the tank is stirred frequently to avoid sedimentation. Manufacturers of commercial slurries, which have different sedimentation qualities, usually have reliable information on the requirements to keep the particles in suspension. The mechanical energy generated by stirring the slurry is converted to heat, which is generally enough to keep the liquid from freezing during periods of low temperatures.

Membrane pumps are well suited to transport the slurry from the storage tank to the mixing unit next to the storage tank, where it mixes with the inlet water. These types of pumps are accurate, dependable, and inexpensive; most are automatically controlled. The maximum dosage rate for a large slurry doser (storage volume 80 m^3) is 2 tonnes/hr.

A power failure results in a complete shutdown of the doser, creating sedimentation in the slurry storage tank and making it difficult to restart the system. A back-up battery system is often installed to help prevent such problems. This system ensures that the water quality in the stream is maintained at a reasonable, but not optimal, level.

Fraser et al. (1985) described the Marback, Sweden slurry doser project, in which a wet-slurry doser system operated on the Fyllean River was designed to protect the downstream spawning areas of Atlantic salmon. The liming apparatus consisted of an 80-m^3, insulated, concrete holding tank containing a wet suspension of limestone particles with an average diameter of 2 μm, mixed with a dispersing agent. The suspended slurry was then released in 10-min pulses through an insulated pipe into the outflow waters at the base of a small hydroelectric station on the river or stream. The dose is determined by a microprocessor programmed to estimate pH from streamflow.

In April 1984, when the Fyllean River flow rate was 10.2 m^3/sec, the slurry doser was dispersing the base material at a rate of 120 kg/hr (3 g/m^3). Alkalinity and pH were 30 μeq/L and 6.8 at a point 3 km downstream from the treatment site. Ever since liming began in Marback and at the Ryaberg site 15 km upstream, Atlantic salmon have spawned successfully.

The Maryland Department of Natural Resources, in cooperation with LLI, has operated electrically powered wet-slurry dosers on two Coastal Plain streams where several anadromous fish species spawn (Dalpra, 1988; Janicki and Greening, 1988a,b; Greening et al., in press). Mattawoman Creek in Charles County (Figure 2.22) and Bacon Ridge Branch in Ann Arundel County were selected for experimental treatment during storm episodes. An on-site computer continuously monitors streamflow. On the basis of flow and previously collected water chemistry data, the computer estimates stream acidity and pumps an appropriate amount of slurry from the storage silo to a mixing tank. The slurry is then diluted with stream water and pumped to the stream. For the first 2 years, the dosers were effective under most conditions. During some storms, however, stream dosing was interrupted by clogging of discharge pipes or power failure. The dissolution efficiency of the limestone materials was reported to be high.

Figure 2.22 Electrically powered wet-slurry doser on Mattawoman Creek in Charles County, Maryland. (Photograph courtesy of Ronald J. Klauda.)

Battery-Powered Dosers. Battery-powered dosers are generally small, portable, and easily installed devices, typically powered by a single 12-V automobile battery (Figure 2.23). Larger versions powered by multiple 12-V batteries are also commercially available. Fraser et al. (1985) described these dosers as essentially modified vibrating fish feeders, used in many fish hatcheries. The dosers, manually or pneumatically filled with dry-powdered limestone, are typically installed on or near a bridge or on a rack over a weir in the stream. The limestone-dispensing devices are multiple, cone-shaped storage and feeder units at the bottom of a limestone storage bin and are attached to a round vibrating plate mounted a few centimeters below the mouth of the cone. Vibrations cause the powdered limestone to fall through a slit at the bottom of the storage and feeder unit, onto a vibrating plate from which it is placed into the stream. The application rate can be controlled automatically by microprocessors. Battery-operated dosers in Sweden are equipped with 1–15 vibrator units and limestone storage bins of 0.1–30 m^3. The standard model is equipped with three to five vibrator units.

Fraser et al. (1985) described some of the major design features

Figure 2.23 Small battery-powered doser in Sweden. (Photograph courtesy of James Fraser.)

and limitations for these dosers. The exterior of the storage bin is usually made of metal, but may be fiberglass or some other weather-resistant material. The cone-shaped storage and feeder units on existing models are textile (polyester) bags, but metals or plastics are also used. Battery storage containers are insulated to protect and extend the life of the battery during subfreezing weather.

The dosers generally use limestone powder with an average particle diameter of 12 μm. The smaller models (storage volumes of 0.1–5 m^3) can store 0.12–6 tonnes of limestone. Several small dosers are preferred over a single large unit, because powder compaction and blocking of the vibrators increase with the size of the storage unit.

The use of standard dosers with three to five vibrating units is limited to streams with maximum flows of 1–2 m^3/sec. The largest commercial battery-powered dosers (15 vibrator units), which have dosage capacities of 10 g/m^3, can treat maximum flows of 8 m^3/sec. Each limestone storage and feeder unit and attached vibrating plate can dispense limestone at a rate ranging from 0.2 to 20 kg/hr. The model equipped with 15 vibrator dosing units thus has a maximum capacity of 300 kg/hr.

The dosage rate may be fixed or adjusted to treat variations in flow. Float (mercury) switches (or remote flow sensors) on a staff gauge are used to relay information to a microprocessor or automatic control device to regulate the number of vibrators dispensing the base material. The quantity released per vibration can be manually adjusted.

All electronic components are powered by 12-V batteries. The batteries function from 6 weeks to 6 months without recharging, depending on the number of vibrator units that are operating and the ambient air temperature. Typically, one battery powers up to four vibrator units for 6 months. Because the system is battery powered, relatively remote sites can be treated. Operation of the system may be hindered, however, by humid conditions or extended periods of below-freezing weather. Shutdown may occur during extended periods of temperatures below -5° C because the lower end of the limestone storage and feeder unit and the attached vibrating plate cannot be enclosed for protection from the weather.

The storage bins of small installations are usually filled manually with bagged limestone. The larger systems can be designed to be filled pneumatically. At maximum flow rates the doser typically must be refilled every 24–48 hr. At average or low flow, the units require refilling weekly or less often. A battery-operated doser ("Borlange limer") was used in the Brodalsbacken area of southern Sweden to maintain favorable pH and low inorganic aluminum concentrations in downstream spawning areas. The doser was situated over a weir on a small tributary stream (Sandvadsbacken). Dry-powdered limestone with an average particle diameter of 12 μm was stored in a 650-kg capacity bin and dispensed by three vibrator units. At low stream (0.04 m^3/sec or less), only one vibrator operated. The second vibrator was activated to dispense powder from the storage unit when the flow increased to about 0.07 m^3/sec and the third vibrator at flows of 0.1 m^3/sec or greater. In October 1983, the doser was dispensing limestone at the rate of 0.11 kg/hr (3.3 g/m^3) into a streamflow of 0.01 m^3/sec.

Water-Powered Dosers. In water-powered dosers, regulated streamflow is used to control the dispersion of limestone. In the only water-powered system that has been operated for several years, a turntable and tipping bucket are located below the storage bin. The "Boxholm" doser is a newer device that operates on a smaller principle. Another system, placed in operation at U.S. Fish and Wildlife Service liming sites in Massachusetts and Tennessee, uses an apparatus similar to a paddle wheel that turns an auger. Water-powered systems are typically used in small, moderately acidic streams that do not have large variations in flow and that are not near a source of electric power.

The "Boxholm" doser has a limestone storage bin with a small opening in the bottom, a conveyor belt, a rocker arm with a hammer mechanism at one end, and a tipping bucket at the other end (Figure 2.24). A photograph of this type of water-powered doser system installed in 1989 in a small stream in Pennsylvania is shown in Figure 2.25.

A small portion of the streamflow, manually regulated by a valve, is diverted through a pipe or trough into the bucket attached to the rocker arm. As the bucket fills with water, a continuous flow of limestone is gravity fed to a conveyor belt located a few centimeters below the mouth of the limestone storage bin. When the bucket is filled with the stream water, it tips over, causing the conveyor belt to spill limestone into the bucket. The movement of the bucket also causes the hammer mechanism to strike the side of the limestone storage bins. The resulting vibration loosens any limestone powder that has become lodged onto the sides of the bin.

Water-powered dosers are installed either in or adjacent to the stream. The doser is often located so that the material is dispensed into a turbulent section of the stream to enhance the dissolution efficiency of the dry particles. Construction of storage bins with capacities exceeding about 40 m^3 is not feasible because the striking force of the hammer mechanism is then ineffective. Other

Figure 2.24 Drawing of the "Boxholm" water-powered doser. (Source: Boxholmkonsult AB, 1989.)

specific design criteria were outlined by Fraser et al. (1985) as follows:

> The dosers are typically constructed of metal with a metal or polyvinyl chloride (PVC) influent water pipe. A concrete or other similarly robust foundation is necessary for the system to withstand increased water flow and ice build-up. The foundation must also be level in order for the rocker arm/tipping bucket mechanism to function properly. The dosers are de-

Figure 2.25 "Boxholm" water-powered doser operation in Pennsylvania. (Photograph courtesy of Dean Arnold.)

signed for streams having flow rates up to, but not greater than, 3 m^3/sec. This maximum flow rate is based on a limestone dosage rate of 10 g/m^3. Additionally, at least a 1-m drop in head height is required for the apparatus to operate.

The dosers are typically equipped with tapered storage tanks having volumes of up to 40 m^3.

The dosers have a maximum dosage rate capacity of 100 kg/hr or 10 g/m^3. The dose is controlled by manually adjusting the width of the opening at the bottom of the storage bin, as well as by varying the flow of water to the bucket.

The storage bin is filled manually with bagged limestone or pneumatically from a small, high-pressure tank truck. During high flow periods it may be necessary to refill the smallest storage bins every 6–12 hr. During most of the year (flows less than 1 m^3/sec), the bins typically need to be refilled weekly. This type of doser is limited to areas that do not have below-freezing conditions for extended periods of time. Operation of the rocker arm mechanism is somewhat hindered by frost conditions.

Water-powered dosers have been effective in improving water quality in several marginally to moderately acidic streams in southwestern Sweden. One such device was installed on Silkbacken, near the community of Hallefors (Fraser et al., 1985). During April 1984, when streamflow was 0.54 m^3/sec, the doser operated at selected dosing time intervals, and therefore different dosages, dispensing powdered limestone with an average particle diameter of 14 μm. At a 5.7-min dosing interval and a dosage rate of 4.0 kg/hr (2.1 g/m^3), alkalinity did not change; pH increased from 6.1 upstream from the doser to 6.2 at a point 50 m downstream. Decreasing the dosing interval to 2.3 min and increasing the dosage rate to 9.9 kg/hr (5.1 g/m^3) changed alkalinity and pH from 70 μeq/L and 6.1 upstream from the treatment site to 100 μeq/L and 6.8 at a point 300 m downstream.

The system, once commercially available in Sweden, was named the *Hallefors Limer* by one manufacturer after the town in which it was first installed (Cementa Movab, 1989). The effectiveness of another Hallefors limer was investigated in April 1984. This system used limestone powder ≤500-μm diameter to treat a flow of 0.94 m^3/sec. Limestone applied at a rate of 22.5 kg/hr (5.5 g/m^3) increased alkalinity and pH from less than 10 μeq/L and 4.7 upstream from the doser to 20 μeq/L and 5.6 at a point 1 km downstream. Although Hallefors limers are no longer commercially available in Sweden, 28 were in operation in 1990 (E. Thornelof, personal communication, 1990).

Diversion Well

A diversion well is a container filled with base material and located along a stream bank or in the bottom of a stream bed. Although a variety of design options are available, the most commonly used wells are cylindrical structures filled with base material—usually crushed limestone 6–8 mm in diameter (Figure 2.26).

The apparatus receives all of its energy from the water, which is diverted from the stream through a tube that opens at the bottom of the well. The diverted water flows upward through the crushed limestone. The upwelling agitates the limestone, creating a fluidized bed that results in abrasion and dissolution of small alkaline particles. The diverted stream water with increased ANC then escapes over the lip of the well or through an outlet pipe. Increasing the diameter of the top of the well decreases the fluid velocity at the outlet, and undissolved particles sink back into the well rather than being entrained in the outflow.

The process typically results in the use of 80–90% of the base material, although the percentage may vary between 60% and 90% depending on the well design. These utilization efficiencies are typical of operating systems in streams with initial pH values in the range of 4.9–6.5 (Fraser et al., 1985).

The typical maximum streamflow neutralized by the early diversion wells was about 0.2 m^3/sec. Recent design modifications and the construction of two or more wells in parallel, with stream water diverted to the additional well or wells during especially high flows, may greatly improve this technique and allow treatment of flows greater than 1.0 m^3/sec. A two-well system operating since 1980 at Piggaboda in Smalland, Sweden increased stream pH by one unit; the limestone in the well dissolved at the rate of about 5% per day (Figure 2.27).

In contrast to other systems for neutralizing flowing waters, diversion wells are not assembled from several rather independently operating parts, but rather are designed as integrated units for specific sites. The design process relies on a substantial knowledge of chemical engineering and fluidized bed reactors. Raser et al. (1985) described design criteria for a fluidized diversion well as follows.

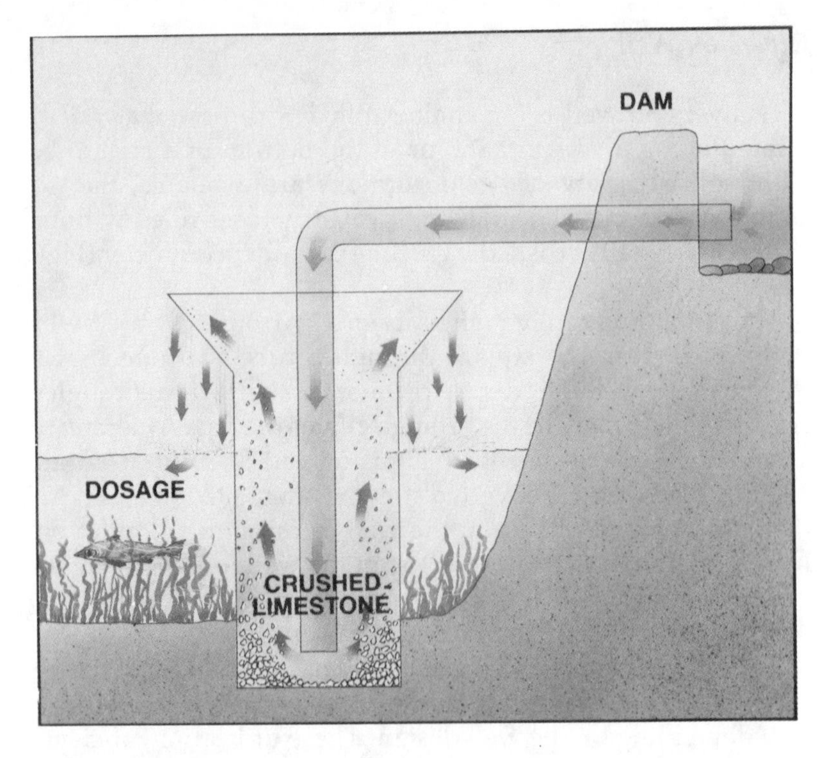

Figure 2.26 Diagram of a diversion well. (Source: Cementa Movab, 1989.)

A diversion well is normally constructed of stacked concrete rings over a concrete foundation. The tapered tip can be made of steel or reinforced fiberglass. The wells use limestone gravel with average absolute particle diameters of 4–9 mm.

The application of single diversion wells is generally limited to first-order streams with average flows ≤1.0 m^3/sec. At flows slightly greater than 1.0 m^3/sec, at least two diversion wells in parallel are recommended to provide adequate neutralization. Sufficient pressure to fluidize the limestone bed requires a hydrological head of greater than 1.3 m. A small concrete or metal dam is usually required upstream from the well to provide the necessary control for the intake structure.

The first step in the design process involves determining the amount of limestone required for a daily dose, which is based on the flow rate and neutralization demand of the stream. Daily dose is calculated by titrating a sample of stream water with limestone and multiplying the result by the daily flow rate. To calculate the

Figure 2.27 Two-well diversion well system operating in Sweden. (Photograph courtesy of Cementa Movab.)

total amount of limestone needed in the well, one divides the amount of limestone required in a daily dose by the percent of the total amount of limestone dissolved in the well per day. The amount dissolved in the well per day can be determined by knowing the inlet velocity and the diameter of the well, and computing the linear upward velocity of the water in the well.

A single diversion well designed to treat a maximum flow of about 1.0 m^3/sec typically is refilled weekly. All existing wells are filled manually.

Fluidized diversion wells have been used to neutralize small streams in Scandinavia since the late 1970s. Bernhoff (1979) reported the use of "limestone wells", consisting of limestone-filled concrete rings embedded either in the stream bank or in the bottom of the stream to release limestone into flowing water in Norway and Sweden. A similar diversion well system was recently installed in a Pennsylvania stream (R.W. Brocksen, personal communication, 1990). The Swedish Ministry of Agriculture Environment Committee (1982) described these wells as uncomplicated and reliable devices that are effective in the liming of running waters.

Limestone Rotary Drum

Rotary drums are cylindrical containers filled with limestone aggregate, powered by water diverted from the stream and directed across a sluiceway. Openings in the bottom of the sluice are located directly above each drum. As water falls through the openings, blades attached to the exteriors of the drums begin rotating, as in a waterwheel. Limestone aggregate is either manually loaded into each drum or automatically fed to the drums by the action of a reciprocating feeder at the bottom of a hopper. In these self-feeding drum systems, a flexible drive shaft is used to convey power directly from the rotating drum shaft to the reciprocating feeder (Figure 2.28).

Water volume through the sluiceway determines the speed at which the drums rotate and, consequently, the amount of aggregate supplied to the drum and ultimately to the stream. Thus, within the capacity of the drum, the amount of limestone needed to maintain a target pH/alkalinity value is available, regardless of flow conditions. This dosage control principle is similar to that for the tipping bucket mechanisms in the water-powered dry-powder system.

The grinding of the limestone aggregate within the drum is a wet autogenous process (i.e., the aggregate itself is used as the abrasive agent). The aggregate is mixed with sluice water entering the drums through tiny holes in the drum exterior. Drum rotation causes abrasion, which in turn produces "fines" (90% of the particles are less than 30 μm in diameter) that are released into the stream through the same holes in the drum exterior.

Recent developments include the use of screens with various mesh sizes to control particle size. Output of the produced fines is controlled by aggregate size and the rotational speed of the drum. Several drums can be operated in parallel so that increased water flow causes the rotation of additional drums. Both mechanically powered (Lovell, 1973) and water-powered (Zurbuch, 1963) rotary drums loaded with limestone aggregate were used successfully to neutralize streams acidified by drainage from coal mines. Lovell (1973) experimented with a prototype electrically powered rotary drum, but the process was never applied on a continuous basis.

Two stations at which the water-powered rotary drum method was used were operated as demonstration projects in

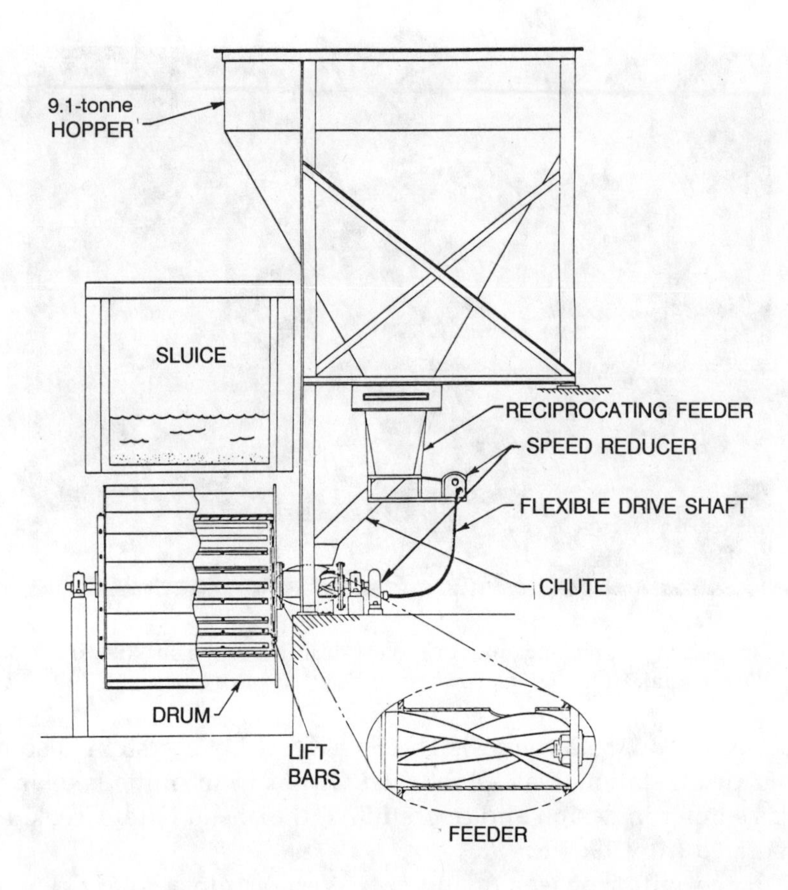

Figure 2.28 Diagram of a self-feeding limestone rotary drum. (Modified from Gencsoy et al., 1983.)

West Virginia (Zurbuch, 1984). One station was installed in 1963 near the headwaters of Otter Creek to establish and maintain an indigenous population of brook trout throughout 16 km of the stream (Figure 2.29). The second station was installed on a tributary stream of Sherwood Lake, a softwater lake in an attempt to increase fish production.

Four water-powered rotary drums, partly filled with high-grade limestone aggregate (>95% $CaCO_3$ by weight, mean absolute diameter of 38 mm) were installed at the Otter Creek site. A water flow of 0.014 m^3/sec was needed to rotate the drums; an operational flow of 0.113m^3/sec produced a limestone dose of 5.4 kg/hr. Limestone treatment began in November 1964 and continued until 1979. From 1965 to 1969, the quantity of limestone distributed into the stream annually ranged from 41.2 to 68.8

Figure 2.29 Self-feeding limestone rotary drum operating on Otter Creek, West Virginia. (Photograph courtesy of Peter Zurbuch.)

tonnes (mean, 53.2 tonnes). The Sherwood Lake system, also a four-drum station, was established within 1 km of the lake and was similar in design and operation to the one at Otter Creek. It operated from 1964 to 1969.

In the initial phases of the two West Virginia rotary drum stations, each drum was manually filled with enough limestone for about 4 days of operation. Experiments with the drums and with stone 38 and 50 mm in diameter indicated that the grinding efficiency and output of fines were maximized when the drums were only party filled (25% of the total volume). Thus, to maintain the optimal limestone mass, a system was designed to carry limestone from a storage area to the drum automatically.

A self-feeding rotary drum system, installed in December 1988 on Dogway Fork in West Virginia, was still operating in early 1990. The main design features of the self-feeding rotary drum system include a hopper, a feeding mechanism that varies with the speed of the drum, and a method of metering and transporting limestone aggregates from the storage hopper into the interior of a water-powered rotary drum. Zurbuch (1984) described the design criteria of this system as follows:

The rotary drum, waterwheel blades, storage hopper, and building to house the system are constructed of metal. The sluiceway is constructed of wood.

Limestone aggregate, approximately 38 mm in absolute diameter, is the preferred neutralizing agent. Annual supply is stored both inside and outside the building that houses the drum system.

A single, self-feeding rotary drum can neutralize acidic stream flows of approximately 1.0 m^3/sec for a distance of up to 15 km downstream from the treatment site. A series of drums, therefore, can treat much greater stream flows (probably in excess of 10 m^3/sec).

A small dam is necessary to create a drop in stream head of greater than 1.5 m from the sluiceway intake to the drum outlet and also to control the water diverted across the sluiceway.

Limestone aggregate can be fed continuously into the interior of the drum at the rate of 54 kg/hr. Therefore, at the optimal water flow of 0.113 m^3/sec in the sluiceway, a single drum can produce a limestone dose of 15 g/sec or 133 g/m^3. A calibrated opening in the reciprocating feeder mechanism acts as a metering device to control the amount of aggregate fed to the drum during each stroke of the feeder. When the flow rate is higher, the drum turns faster, the number of strokes of the reciprocating feeder per unit time increases, and more limestone is supplied to the drum.

Bulk delivery of limestone can be taken annually and the limestone can be stockpiled at the site. The 9.1-tonne storage hopper for each self-feeding drum requires only weekly refilling from the reserves (usually with a front-end loader).

Limestone Barrier

In Sweden, Canada, and the United States, attempts have been made to increase stream water pH by placing gravel-sized limestone into streams. The term *barrier* is somewhat misleading because the water is not prevented from flowing. The limestone is actually a filter to water flow. Most of the early experiments with limestone barriers for stream neutralization were conducted to treat streams acidified by acid mine drainage. Barriers are held in place either by larger stones on upstream and downstream sides or by commercially available wire gabions (Figure 2.30). In some situations, limestone is placed directly on the stream bottom. Most of the results of about 20 years of experience with barrier installations have been disappointing. These systems have rarely been observed to meet water quality targets at varying flows. Because no real design information exists, the following discussion of project successes and failures serves to illustrate the different approaches and their effectiveness.

In 1970 and 1971, six limestone barriers were constructed in streams in the Trough Creek watershed, Huntingdon County, Pennsylvania, that were impacted by drainage from nearby coal mines (Pearson and McDonnell, 1975a,b; Yocum, 1976). The barriers consisted of cores of crushed limestone placed in the stream channel, constrained by heavy rocks on the upstream and downstream faces. The quantity of limestone in the barriers ranged from 130 to 830 m^3 (Yocum, 1976).

Pearson and McDonnell (1975a,b) and Yocum (1976) reported that, during periods of low flow, performance of the six limestone barriers was excellent in reducing acidity and increasing the pH of the water. The effectiveness of the barriers, however, was only marginal at normal flows and nil during high runoff periods. Barriers were most efficient in neutralizing waters at initial installation or after reconditioning. Deterioration of barrier performance was due to two major factors: (1) accumulation of silt or iron sludges that clogged the barrier and (2) coating of the stone surfaces with silt or iron hydroxide precipitates. Yocum (1976) recommended that in-stream neutralization with limestone barriers be limited to waters having very low iron concentrations (less than 1 mg/L) and that a maintenance program be established to recondition (scrape and clean) the barriers periodically to restore neutralization efficiency.

Leedy and Franklin (1981) reported that limestone barriers were only marginally successful in treating streams affected by acid mine drainage. Furthermore, their use sometimes resulted in the formation of floc that could be detrimental to the habitat of invertebrates on the stream bottom. Limestone barriers also have been used to neutralize streams that are less strongly acidic than

Figure 2.30 Limestone barrier operating on Gifford Run in Pennsylvania. (Photograph courtesy of Dean Arnold.)

those impacted by acid mine drainage, although similar problems were reported for treatment of these streams. Several more recent case studies are described below.

An experimental limestone barrier in Constable Creek, an acidic stream (pH 4.8) that flows into Big Moose Lake, in the Adirondack region of New York was constructed with 55-gal drums, perforated at both ends, and stacked 2 barrels high across the width of the stream (Driscoll et al., 1982). About 2.6 tonnes of crushed limestone (mean diameter 10 mm) was used to fill the barrels and the spaces between them. Trash screens were placed upstream to prevent clogging of the pores of the limestone barrier. Like the other barriers, however, this barrier neutralized the stream for only about 1 week, after which coatings by silt and metal hydroxides prevented further neutralization.

Arnold (1981) reported on the effectiveness of an experimental limestone flow-through device for increasing the pH of Gifford Run, a small stream in Pennsylvania (Figure 2.31). During fall 1976, about 36 tonnes of limestone, secured by wire gabions, was placed in the stream. On the first day, the pH rose nearly one unit. Thereafter, increases in pH were only 0.1–0.4 unit during occasional low-flow periods. It was concluded that greater increases

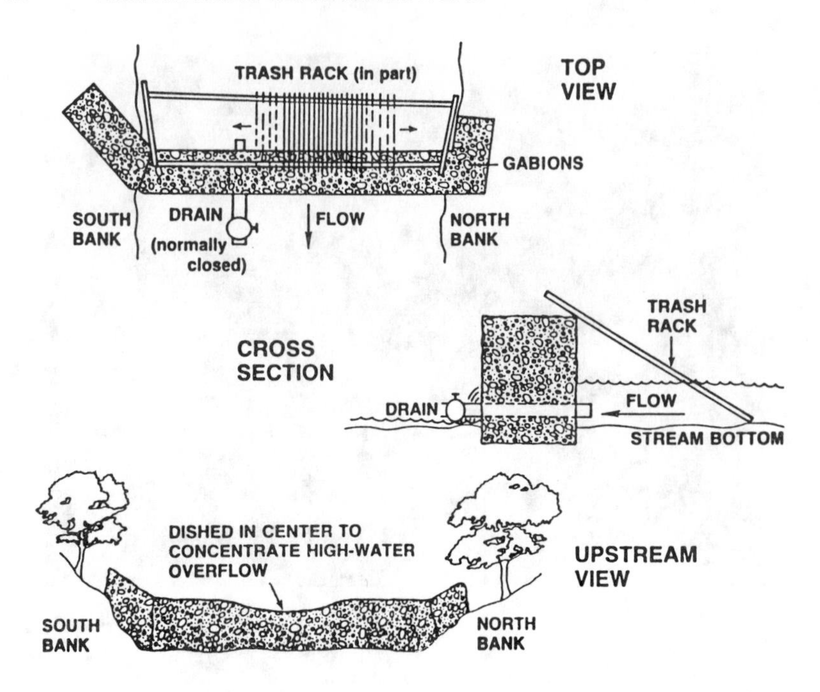

Figure 2.31 Diagram of limestone barrier on Gifford Run in Pennsylvania. (Modified from Arnold et al., 1988.)

would require that a series of these barriers, each containing high-calcium limestone of uniform size (100 mm in diameter), would be needed. Also, trash rack devices (rows of vertical bars about 5 cm apart) upstream from the barrier were recommended to prevent clogging by leaves.

In more recent experiments at Gifford Run, a limestone barrier was constructed in October 1986 with a series of commonly used commercial wire gabion units (1 x 1 x 2 m) joined to increase strength of the installation (Arnold et al., 1988). The uniform stone size (recommended above) was not available. Although the fine particles used in the barrier resulted in initial pH increases of about 0.5 unit, they dissolved within about 2 weeks and net treatment effects decreased rapidly. For the remaining 2 months of operation, pH and alkalinity changed little. Another problem encountered was the accumulation of leaves on the barrier, despite the addition of a trash rack (consisting of sloping bars parallel to the streamflow) that was intended to direct leaves over the barrier. The investigator remained optimistic, however, about the ability of properly installed limestone barriers to effectively treat mildly acidic streams in remote areas.

Fraser et al. (1982) reported on the unsuccessful use of limestone (about 20 mm in diameter) at one site in Sweden, where the barrier did not remain in place. High flows (up to 125 m^3/sec) washed much of the material downstream into a lake, forming a delta. The formation of the delta had little effect on the lake pH, because the decreased stream velocity over the delta resulted in little abrasion of the limestone gravel.

Another site in the Hogvadsan River system was treated with 100 tonnes of limestone over a 300-m reach to a depth of 50 cm. Little change in pH resulted from this treatment. For example, during June 1981, the pH was 5.2 upstream and 5.4 downstream from the barrier (Fraser et al., 1982).

Limestone barriers have been built at several sites in Canada. In Killarney Provincial Park, Ontario, a 100-m stream reach was filled to a depth of 1.0–1.5 m with 318 tonnes of dolomitic agriculture limestone (about 12–18 mm in diameter); 27 tonnes of much larger limestone was dumped at the downstream end to prevent the smaller pieces from being washed away. During summer low flow, the barrier raised the pH (relative to that during spring high flow) to acceptable levels (Fraser et al., 1982). The barrier reportedly added about 18 kg of $CaCO_3$ alkalinity per year to the stream. Fish were reportedly able to survive downstream, but not upstream, from the barrier.

Experimental placement of limestone rock into a stream bed was attempted in Lower Great Brook, Nova Scotia, Canada (Fraser et al., 1982). In August 1980, 9 tonnes each of calcitic and dolomitic limestone (about 50–150 mm in diameter) were dumped from a truck and spread with a backhoe in a single layer, 3–4 m wide, across the bottom of the stream where the width was 7–8 m. This particular example does not illustrate a true barrier, however, as water about 0.5 m deep was flowing over the limestone. Although an initial increase in pH was detected, the limestone eventually became coated with an insoluble metal precipitate and neutralization was greatly reduced.

Watt et al. (1984) reported the results of the construction of limestone barriers at six sites in Nova Scotia. About 100–300 tonnes of limestone gravel was placed in these streams. The pH increases observed during these experiments ranged from nil during maximum discharge in winter to 1.6 pH units during low flows in summer. Difficulties in using this technique in Nova Scotia streams included tendencies of the gravel to (1) be deactivated by chemical and biological coatings, (2) become covered by sediments if placed in an area where current was too low, and (3) be washed downstream if placed in an area where current was too high. A suggestion to match limestone gravel size with that of the substrate present in the area where the barrier was to be placed was considered impractical, however, for barrier construction in large streams or rivers.

Techniques Used in Watershed Applications

Application techniques for liming soils in the watersheds of streams and lakes are similar to methods used in agriculture and for liming lakes. Although actual application methods are not particularly difficult, the amount and location of applied material are critical.

Boat

Wetlands can be limed by using any of the boat application techniques described earlier for lakes, provided the boat can be launched and operated in the wetland and the material can be distributed to where it is needed.

Truck or Tractor

Liming of forest soils or wetland areas by truck or tractor also is possible if the techniques described earlier for lakes are used. A watershed in Sweden has been sprayed with powdered agricultural limestone from a tank truck equipped with pressurized blower systems (Figure 2.32). A conventional hydroseeder was used to distribute limestone to portions of the Loch Fleet watershed in Scotland.

Conventional lime-spreading trucks have been used to distribute agricultural limestone or hydrated lime to watershed areas that are accessible by vehicle (Rafaill and Vogel, 1978). In a study reported by Rafaill and Vogel (1978), limestone was applied in steeper areas with an Estes spreader (a blower-impactor device) and hydrated lime was applied to the other areas with a conventional hydroseeder.

Truck application techniques have also been used in the neutralization of acid mine drainage (Fraser and Britt, 1982). Five

Figure 2.32 Truck and blower spreading limestone onto a watershed in Sweden. (Photograph courtesy of Joe Wisniewski.)

extremely acidic lakes (pH ranging from 3.0 to 4.3) located on Peabody Coal Company's River Queen Mine, Muhlenberg County, Kentucky, were neutralized either by the application of agricultural limestone to the watershed (by truck and spreader) or surface water (by truck and blower) of each lake, or by a combination of the two techniques. Four lakes were treated with agricultural limestone at a dosage rate of 44.8 tonnes/ha of treated area within the watershed of each lake. Also, the lake surface and shoreline of all lakes (one exception) were treated with an application of limestone from a truck with a blower attachment. The pH reached 6.0 within 6 months in all lakes receiving direct application, but only after 18 months in the single lake that was not limed directly.

More recently, forest soils have been limed with a truck or tractor in Norway, Sweden, and the United Kingdom (Brown, 1988; Sverdrup and Warfvinge, 1988b). Various land vehicles were used to spread agricultural limestone and pelletized limestone over portions of the watershed.

Helicopters and Fixed-Wing Aircraft

The techniques described earlier for lakes can be used to lime forest soils or wetland areas by helicopter or fixed-wing aircraft. Aerial application may be more feasible in areas where road access is limited. Fixed-wing aircraft may be a logical choice if a suitable airport is near the target watershed. Helicopters were used to apply agricultural limestone and olivine at the rate of 6 tonnes/ha to forested watersheds in Sweden in December 1979. Altogether, 18 tonnes of olivine (comprising mostly magnesium, iron, and silicon) were applied in one watershed and 12 tonnes of agricultural limestone in another (Fraser and Britt, 1982). More recently, helicopters were used to apply limestone over wetland areas in some large liming projects in Sweden. For example, 6555 tonnes were applied to the watershed to reduce aluminum concentrations in the River Morrumsan in southeastern Sweden and 7699 tonnes to the River Issjon watershed in northwestern Sweden to increase pH and to reduce metal concentrations (E. Thornelof, personal communication, 1990).

Helicopters also were used in July 1983 to lime the Tjonnstrond catchment in the county of Telemark, Norway (Rosseland and Hindar, 1988). The watershed contains two small ponds, Lake Ovre Tjonnstrond and Lake Nedre Tjonnstrond. A total of 75 tonnes of powdered limestone was added, corresponding to a watershed dose of 3 tonnes/ha. This dose is about 50 times the amount that would have been used in a direct application to the lakes. The watershed was limed with helicopters because the lakes' water retention times were too short for whole-lake liming to be practical and doser liming of inlet streams was not feasible (one lake had no inlet streams). This liming improved water chemistry in the lakes to levels suitable for supporting brook trout and brown trout. Conditions were still favorable 4 years after the treatment.

The first reported liming of a watershed in the United States was in October 1989. The Woods Lake watershed in the Adirondack region of New York was treated by spreading pelletized limestone from a helicopter (Brocksen et al., 1988; Boutacoff, 1990; Olem et al., 1990) (Figure 2.33).

Timing and Location of Liming

Timing and location of application are important factors in the success of a liming operation or experiment. These considerations are described here.

Figure 2.33 Helicopter liming of the Woods Lake watershed in New York showing pelletized limestone (top inset) and close-up of distribution bucket (bottom inset). (Photograph courtesy of Donald Porcella and Ted Spiegel.)

Timing of Liming

In surface waters that support fish populations, the critical times are periods of high influxes of additional acid, such as those during spring snowmelt and fall rains. For lakes, streams, and watersheds, application should be scheduled to precede these critical events. This is not always possible due to competing factors.

Lakes

For lakes, application immediately preceding critical events is sometimes effective. Lake liming during spring overturn takes advantage of lake circulation patterns to enhance the mixing and distribution of neutralizing agents. However, neutralization may sometimes be effected too late to prevent the mortality of embryos and fry of numerous fish species during the potentially serious pH reductions associated with melting snow (Jeffries et al., 1979; Johannessen et al., 1980; Haines and Schofield, 1980).

Where spring snowmelt is the problem, liming immediately before the melt is of little advantage because meltwater does not mix well with the warmer lake water. Also, the lower lake water temperatures in winter result in inefficient short-term dissolution. Thus, liming for protection from melting snow may best be done in fall. An option that appears to be practical, especially if fish are to be stocked immediately, is to apply the base material during or shortly before fall overturn.

Streams

For streams, spring storms may be a particular cause for concern because of the presence then of fish eggs and larvae. In Maryland for example, two streams are being limed only during spring storms to protect the eggs and larvae of anadromous fishes (Janicki and Greening, 1988a, b).

Watersheds

Liming strategies for watersheds may be influenced by the sequence of frozen ground to snow to snowmelt. If the ground freezes before snow cover develops and remains frozen during snowmelt, spring snowmelt may interact little with recently neutralized soils, and thus preclude the possibility of reducing the acidity of the meltwaters. Bengtsson et al. (1980) indicated that acidified meltwater can be neutralized by overdosing of liming material that is applied directly to the water or by manual application to the snowpack. The approach may not be practical in all situations because limestone fails to mix with meltwater. The method, however, is used in northern Sweden where snowpack and ice are thick (E. Thornelof, personal communication, 1990).

Location of Liming

Distribution of liming material over an entire lake, stream section, or watershed may be the most ideal means of guaranteeing neutralization of the waterbody. This is not always practical, however, because of a lack of time and funds. The overall choice of locations for liming is site specific and technique specific. No firm conclusions have been made, and no specific experiments have been conducted to define the exact factors that influence the choice of location.

Lakes

The application of liming material over an entire lake is not often feasible. As an alternative, limestone can be distributed over the deepest portion of the lake, thus increasing the amount of time for the larger particles of $CaCO_3$ to react while they pass through the water column. This distribution technique is probably best suited for deep lakes having relatively rapid flushing rates.

Placement of limestone in shallow littoral zones may be more effective for the protection of biota, particularly the early life stages (Blake, 1981; Driscoll et al., 1982; Swedish Environmental Protection Board, 1988). Shallow littoral zones are often areas where maximum turbulence caused by wave action and currents enhance the dissolution process. Observations in several Swedish lakes, however, indicated that water movement in the littoral zone neither increased the dissolution rate nor eroded the $CaCO_3$ particles (Sverdrup, 1982b). The biological benefits of liming littoral zones, and not the maximum dissolution rate, should be the primary consideration. For shallow water application, a finer grade of powder compensates for slower dissolution rates and shorter water column retention times of the particles (Sverdrup, 1982b).

Some researchers suggest that applications to areas with sandy bottoms are preferable to applications to muddy bottoms (heavy loams or clays); sandy bottoms reduce the loss of limestone by chemical bonding with organic sediments (Boyd, 1982a; Driscoll et al., 1982).

If lake size or economics limit a liming program to a small portion of a lake (e.g., a shallow cove), fish survivability may still be enhanced. Hall et al. (1980) and Muniz and Leivestad (1980) indicated that certain fish species seek "havens" or refugia during incidents of episodic pH reductions. The probability of finding refugia with elevated ANC increases with lake size and volume, regardless of the fact that increased volume, by dilution, dampens fluctuations in water chemistry (Muniz and Leivestad, 1980). Liming to create less acidic (or neutral) havens has been conducted successfully in Scandinavia (Fraser et al., 1982).

Another example of a treatment technique for specific surface waters is the application of limestone gravel directly to spawning sites and areas of spring upwelling. Limestone materials can also be placed along lake shorelines (Wright, 1984, 1985; Booth et al., 1987; Rosseland and Hindar, 1988).

Streams

The most important concern for location of stream liming devices is their placement at some point above the target area to provide a zone where metals may precipitate. Large quantities of aluminum often precipitate on the stream bottom some distance downstream from liming because the solubility of the metal is reduced (Driscoll et al., 1982; Dickson, 1983). The precipitated aluminum may be harmful to fish and other aquatic biota. For this reason, the Swedish Environmental Protection Board (1988) recommended that the dosing point be at least 200 m upstream from the target zone. Of course, not all streams that are candidates for liming have concentrations of metals that precipitate and negatively affect the biota.

Watersheds

Experience with watershed liming has indicated that it is very important to apply the base material to major water pathways. This practice reduces the area treated and the amount of base material required. In the most successful projects, limestone was applied directly to discharge areas or wetlands. Treatment only of recharge areas, i.e., dry soils, has been only marginally successful (Lessmark, 1987). In summarizing the results of 52 soil liming projects, Lessmark (1987) reported discouraging results for treatments where limestone was applied only to recharge areas. Only soil liming that included discharge areas effectively neutralized the surface waters. Tervet and Harriman (1988) observed little neutralization of Loch Dee in Scotland during early treatments of the watershed, but later liming, directed at major water pathways and headwater streams, resulted in more substantial increase in pH of the lake.

Methods for Determining Dosage and Reacidification Rates

The methods available for determining the dosages of base to be applied and the various models that have been developed to determine rates of reacidification of treated systems are next discussed. Examples of calculations for a limestone dosage and reacidification model for surface waters were presented by Fraser et al. (1985). In Appendix B, selected modeling approaches are applied for estimating the limestone dose and time required for reacidification to treat diverse lake situations.

Dosage Calculations

Dose calculation methods described in the literature vary widely in complexity. During the early years of liming, dose calculations were often only rough approximations based on acid-base titrations. Now, rule-of-thumb calculations based on many years of liming experience are often adequate for determining material doses.

The amount of alkaline material required varies according to the following factors:

- where material is being applied
- type of material used
- material dissolution rate
- volume of the water body (lakes)
- flushing rate of the water body (lakes)
- flow (streams)
- base saturation of soil (watersheds)
- acidity of the water
- acidity and make up of associated sediments
- amount of organic material in the water
- acid input from incoming waters
- acid input from atmospheric deposition

Recent methods developed have been based on these conditions. The methods use information on dissolution kinetics, chemical equilibria, and particle settling velocities. A single rule-of-thumb dosage rate representing the quantity of liming materials needed to neutralize the "average" acidic surface water is difficult to provide because of the above factors. As judged from several years of data on required dosages (particularly for powdered agricultural limestone used in North American and Scandinavian liming projects), yearly dosages of about 0.2–2 tonnes/ha may be a reasonable average range if the limestone is applied directly to the lake surface and the lake retention time is about 1 year (Fraser and Britt, 1982; Fraser et al., 1985; LLI, 1987). A one-time dose of about 5 tonnes/ha may be typical if the limestone is applied to the watershed (Brown, 1988).

Lakes

Various dose calculation procedures developed for lakes are described here and summarized in Table 2.9.

Swedish Dose Calculation Model. The dose calculation method presented in the guidelines for liming lakes in Sweden is based on 10 years of field experience (Swedish Environmental Protection Board). In this method a simple model is used that relates pH and alkalinity of the lake and the lake retention time to the recommended dose in grams per cubic meter. The result is the dose of limestone required to provide immediate neutralization and protection from continued acidic deposition over the life of the project. This duration is calculated by using a diagram that relates lake retention time to the duration of liming. This dose calculation model does not account for differences in lake depth or particle size distribution of the limestone. Although the method is empirical, it appears to result in good overall estimates of limestone doses to lake surfaces.

Lake Hovvatn Model. Wright (1984, 1985) used a simple model based on a titration curve to estimate the dosage required to neutralize acidify, increase alkalinity, and precipitate aluminum in Lake Hovvatn, an acidified lake in Norway. In the model, it was assumed that limestone dissolution efficiency was 50%. Sverdrup and Warfvinge (1988b) considered the approach acceptable, although they recommended using a dissolution efficiency value

Table 2.9 Comparison of Dose Calculation Procedures Available for Lakes

Model	Reference	Principle	Includes Cost Optimization	Advantages	Disadvantages
Swedish dose calculation model	Swedish Environmental Protection Board, 1988	Dose related to lake pH, alkalinity, and retention time	No	Simple, based on several years of empirical data	No dissolution, model does not consider different particle sizes
Lake Hovvatn model	Sverdrup and Warfvinge, 1988b	Titration curve, fixed value for dissolution	No	Simple	No dissolution model
Sverdrup dissolution model	Sverdrup, 1985	Titration curve, dissolution diagram, sediment model	Yes	Field tested, requires few data inputs	Simple sediment model
Norwegian liming handbook	Sverdrup and Warfvinge, 1988b	Titration curve, dissolution diagram	Yes	Simple, field tested, requires few data inputs	Works well only for low doses, no sediment model
DePinto model	Scheffe et al., 1986b	Titration model, dissolution model, aquatic chemical speciation model, sediment model	Yes	Field tested, comprehensive	Requires considerable data input

Table 2.9 Comparison of Dose Calculation Procedures Available for Lakes (continued)

Model	Reference	Principle	Includes Cost Optimization	Advantages	Disadvantages
DeAcid model	Fraser et al., 1985; LLI, 1987	Titration model, dissolution model, sediment model	Yes	Field tested, requires few data inputs, computer-assisted program available	Rigid framework

generated by some other model, rather than one based on a fixed dissolution efficiency.

Sverdrup Dissolution Model. Sverdrup (1985) developed a dose calculation model based on a sinking and dissolving model for limestone powder. Considered in the model are the dissolution kinetics of limestone, a model of the settling of the powder through the water column, and an empirical titration curve of pH and alkalinity. Sverdrup (1985) also estimated sediment dose by using an empirical relation based on field results from several liming projects.

The model is based on consideration of the differential sinking and dissolution of limestone particles (Figure 2.34). When limestone is applied to a lake's surface, the particles are rapidly distributed according to size due to their settling velocity, and consequently they sink and dissolve simultaneously. The larger particles sink more rapidly and thus partly neutralize the water later reached by the smaller particles. This process is represented in the Sverdrup model as a set of differential equations (Sverdrup, 1985).

Sverdrup (1985) considered dissolution to be a very important component of dose calculation. Most limestone powders used in lake liming do not dissolve completely during the application process. The results from calculations with the computerized dissolution model are shown in Figure 2.35 for the dissolution of different limestone powders in 5 m of water with initial pH values of 4–7. It is important to note that these are short-term dissolution results. The dissolution continues years after application at a much slower rate. Long-term dissolution is calculated in a related model of lake reacidification (Sverdrup and Warfvinge, 1985).

A turbulence model was recently incorporated into the dissolution model to further refine the dosage calculation (Sverdrup and Warfvinge, 1988b). This refinement substantially increased the data input requirements without a comparable increase in performance, indicating the unimportance of lake turbulence on dissolution efficiency.

Norwegian Liming Handbook Method. The liming dose calculation procedure outlined in the *Norwegian Liming Handbook* (Sverdrup and Warfvinge, 1988b) was based on the experimental titration curve and the dissolution model included in the Sver-

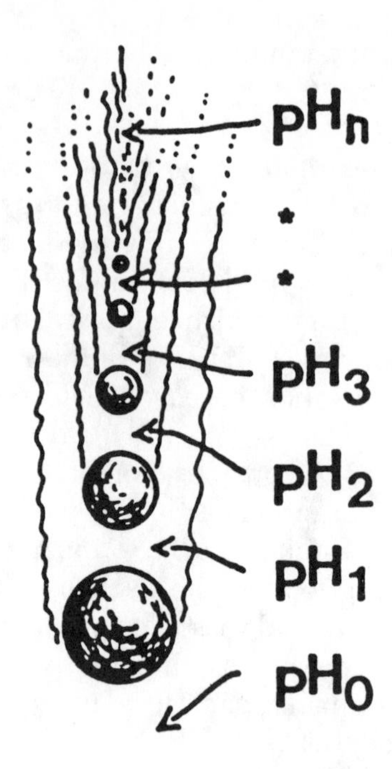

Figure 2.34 Basis for Sverdrup dissolution model. Largest limestone particles sink fastest, partly neutralizing the water below the smaller particles. (Source: Sverdrup, 1985.)

drup model, but did not include the effect of overdosing on dissolution efficiency. The model is therefore most appropriate for determining low dosages in small lakes.

DePinto Model. Scheffe et al. (1986a) described a model developed to determine treatment requirements for lakes to be neutralized with limestone in the Adirondack region of New York. The model, based in many ways on the Sverdrup model, used information on dissolution kinetics and settling velocities and a titration curve model. Factors also included in the model (that were not specifically included in the Sverdrup model) were speciation reactions that may consume alkalinity, such as the speciation of aluminum. The Sverdrup model, however, included these reactions in experimental titration curve results used in the close

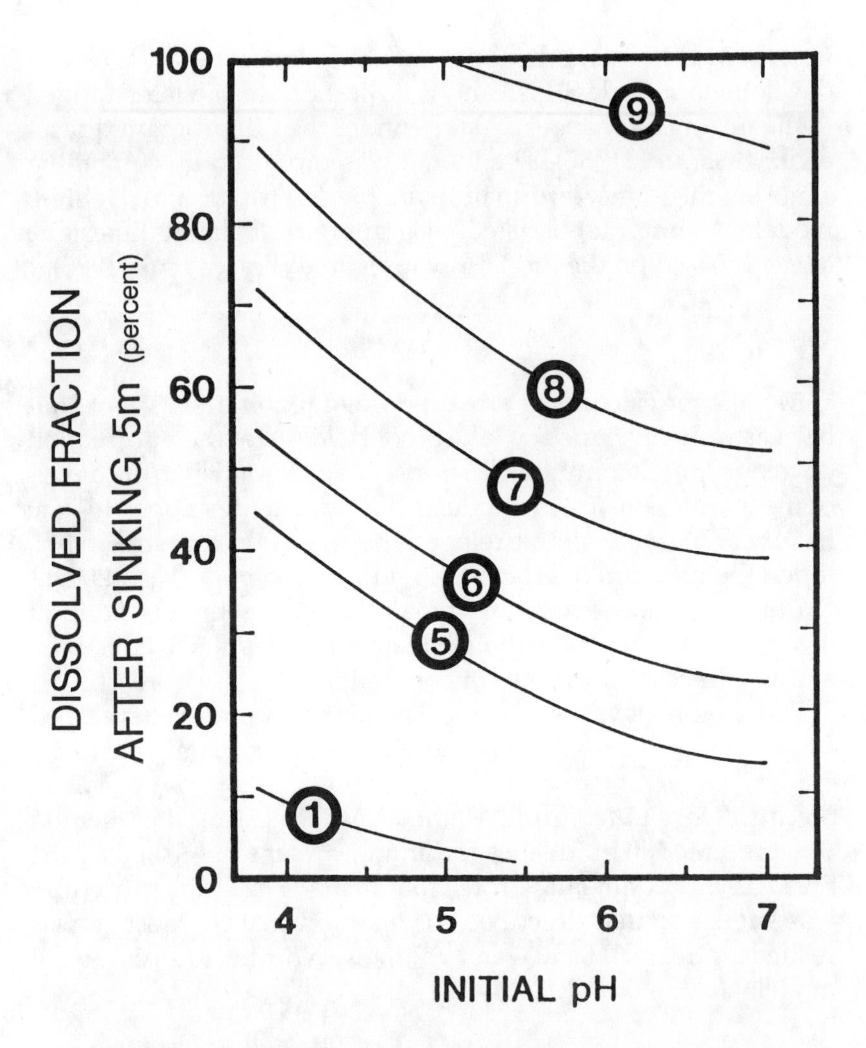

Figure 2.35 Results from the Sverdrup dissolution model showing percent dissolution of different limestone powders in 5 m of water. Refer to Figure 2.6 for description of curve numbers. (Source: Sverdrup, 1985.)

calculation. The model has been verified against laboratory data and applied in the field under the Electric Power Research Institute Lake Acidification Mitigation Program (Young et al., 1989). The model predictions of short-term limestone dissolution efficiency (45%) were lower than observed results (75%), possibly due to the slower settling of limestone particles through the thermocline and to low CO_2 influx. High epilimnion temperatures may also have been a factor.

DeAcid Model. A model for calculating limestone dosages was developed for the U.S. Fish and Wildlife Service and later applied to the treatment of about 27 lakes under the LLI program (Fraser et al., 1985; LLI, 1987). The model was produced by combining some of the same information in the Sverdrup and DePinto models. A computer-assisted program for calculating limestone dosages based on the model was prepared (Fraser et al., 1985).

Streams

Models for calculating stream dosage requirements have not been as well developed as lake models. Much of the information developed for determining lake dosages was applied to the development of stream dosage models. Most models developed to date have been based on simple relations between flow and stream pH, limestone dissolution efficiency, and titration curve results. Unlike one-time lake treatments, the continuous treatments required in stream neutralization allow stream model results to be refined and later incorporated into the operation.

Descriptors of various dose calculation procedures developed for streams are presented here and are summarized in Table 2.10.

Swedish Dose Calculation Method. A dose calculation method was presented in guidelines for liming streams published by the Swedish Environmental Protection Board (1988). The procedure was similar to that described for lakes, in that the dosage was related to stream pH and water volumes according to an empirical model.

Sverdrup Model. Sverdrup (1985) described a stream liming dose model based on a relation between flow and stream pH, an empirical dissolution efficiency diagram similar to the one described previously for lake water dissolution (Figure 2.34) and an empirical or theoretical titration curve. The model was successfully applied in a number of stream liming projects in Sweden (Sverdrup and Warfvinge, 1988b).

DeAcid Model. A model for calculating limestone dosages was developed for the U.S. Fish and Wildlife Service and later applied to the treatment of about 10 streams under the LLI program (Fraser et al., 1985; LLI, 1987). The model, produced by incorpo-

Table 2.10 Comparison of Dose Calculation Procedures Available for Streams

Model	Reference	Principle	Includes Cost Optimization	Advantages	Disadvantages
Swedish dose calculation model	Swedish Environmental Protection Board	Dosage related to stream pH and volume	No	Simple	No dissolution model, no titration curve
Sverdrup model	Sverdrup, 1985	Titration curve, dissolution diagram, dilution calculations	Yes	Simple, field tested, requires few data inputs, based on many years of empirical data	None
DeAcid model	Fraser et al., 1985	Titration model, dissolution model, dilution model	Yes	Field tested, requires few data inputs based on many years of empirical data, computer-assisted program available	None

rating some of the same information used in the Sverdrup model, uses titration curve results and streamflow, pretreatment water chemistry, and limestone particle size to back-calculate the dose required to reach a target pH value (Figure 2.36). A computer-assisted program for calculating limestone dosages based on the model was prepared (Fraser et al., 1985).

Watersheds

Models for calculating dosage requirements for watersheds have only recently been developed. Soil processes are very complex and little experience exists with liming forest soils. Descriptions of various dose calculation procedures developed for soil liming are presented here and are summarized in Table 2.11.

Swedish Dose Calculation Model. Initial guidelines of the Swedish Environmental Protection Board for determining dosages to be applied to a watershed were based on the application of fixed dosages per unit area (Sverdrup and Warfvinge, 1988b). New guidelines divide Sweden geographically into southern, middle, and northern regions, and calculate the dosage based on the percent of the total watershed that is limed (Swedish Environmental Protection Board, 1988). Different dosages are recommended for each region because acidic deposition is more severe in the southern part of the country. It is recommended that between 2 and 10% of the watershed be limed. Also, the limed areas should be discharge areas where at least 50% of the water passes before entering the lake or stream.

Integrated Lake-Watershed Acidification Study Model. Davis and Goldstein (1988) described the application of the Integrated Lake-Watershed Acidification Study model to accept inputs of limestone over selected areas of a watershed. The model was originally developed to quantify acidification by integrating important biogeochemical processes occurring in hydrologic basins. A modification of the computer code allowed for pulsed applications of calcium carbonate.

The model was calibrated and example results were presented for the Woods Lake watershed, New York. No field comparisons were made, though plans were identified for field testing the model during future liming in the watershed. A deficiency of the

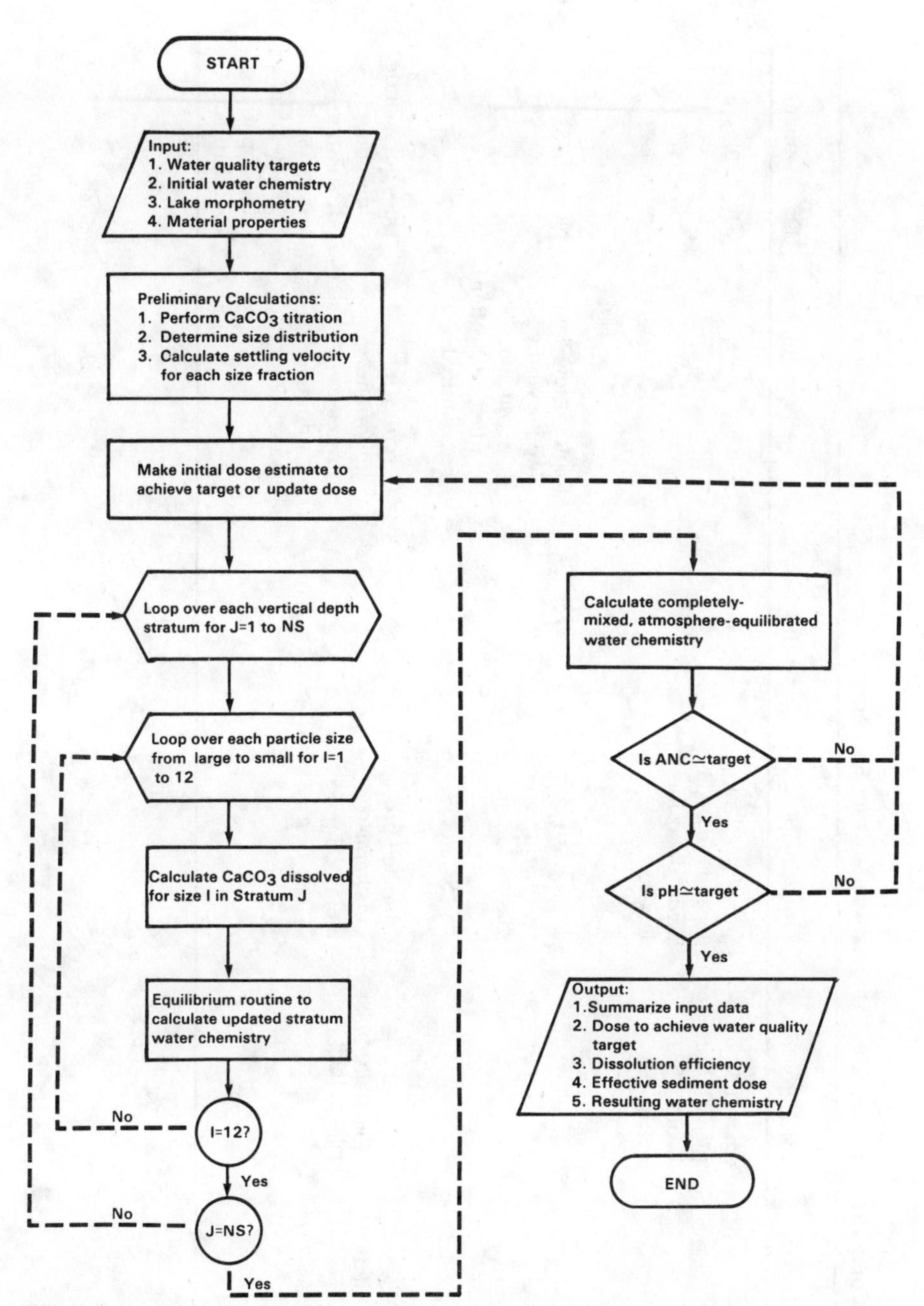

Figure 2.36 Flow chart of the DeAcid model for calculating lake dosage. (Source: Fraser et al., 1985.)

Table 2.11 Comparison of Watershed Models Available for Soil Liming

Model	Reference	Principle	Includes Cost Optimization	Advantages	Disadvantages
Swedish dose calculation model	Swedish Environmental Protection Board, 1988	Dose dependent on location (based on acidic deposition levels)	No	Simple	Only approximate, does not predict response
Warfvinge soil liming model (SLiM)	Warfvinge and Sverdrup, 1988b	Soil chemistry, weathering processes, dissolution model	Yes	Requires few data inputs, verified by experimental studies, calibration not required	None
Integrated Lake and Watershed Acidification Study (ILWAS) model	Davis and Goldstein, 1988; Davis, 1988	Watershed processes, soil chemistry, biological processes	No	Describes important processes within the watershed	Does not include limestone dissolution, requires considerable data input, not yet verified

model noted by Davis (1988) and Sverdrup and Warfvinge (1988b) was the absence of a limestone dissolution algorithm.

Soil Liming Model (SLiM). A model of limestone dissolution in soils was developed by Warfvinge and Sverdrup (1988a,b; 1989) as a dynamic soil model for predicting limestone dosage requirements. The model was shown to take into account several important watershed processes, including the nonequilibrium cation exchange reactions that occur during soil neutralization, an extension of the limestone dissolution models developed for the aqueous environment to conditions appropriate to the soil environment, and information on the hydrologic conditions in the watershed. Because this is essentially a soil model, it includes many watershed processes, including vegetation base cation uptake, nitrogen uptake, silicate mineral weathering, nitrification, aluminum speciation, and evapotranspiration.

Model calculations revealed that the dose must exceed the exchangeable acidity of the soil, that treatments must be applied to at least 50% of the active discharge areas in the watershed at high flow, and that the limestone does not need to be finer than a particle size distribution of 0–1 mm. The model has been applied to liming treatments conducted as part of Swedish, British, and U.S. liming programs.

Predicting Lake Reacidification Rates

Several methods have been developed to predict the rate of reacidification of a lake treated with base materials. Calculation of reacidification rates is important to provide an estimate of when a particular lake must be retreated and to determine whether a particular process or application is economically feasible. Frequent water sampling is a supplemental procedure for determining when to retreat a lake.

The reacidification of a lake involves the same considerations as those described for calculating lake chemical dose, such as the characteristics of the lake water, sediments, and watershed, the severity of acidic deposition inputs, where the material is being applied, and the characteristics of the base material. Reacidification models predict time for reacidification from historical acidic deposition and hydrologic conditions. The quality and quantity of precipitation significantly affect reacidification rates.

Methods developed to date for predicting reacidification rates are presented here and summarized in Table 2.12. Appendix B compares the results of two models for predicting time for reacidification.

Swedish Nomograph

Predictions of the reacidification rates for Swedish lakes are based on an empirical relation with lake retention time (Swedish Environmental Protection Board, 1988). Figure 2.37 shows the duration of neutralized conditions (represented as pH >6.0 or alkalinity >50 μeq/L) for lake retention times up to 10 years. The nomograph is based on a diagram presented by Sverdrup et al. (1985, 1986), but includes data from additional lake liming treatments conducted in Sweden.

Neutralized conditions are predicted by the model to last at least twice the lake retention time, for retention times up to about 4 years. The relation is, of course, not always accurate because the duration of liming is affected by lake chemistry, sediment characteristics, precipitation quality and quantity, and other factors previously mentioned.

Sverdrup Lake Model

Sverdrup and Warfvinge (1985) described a model for predicting time for lake reacidification. The model was based on a hydrologic model, limestone dissolution kinetics, long-term limestone deactivation estimates, and empirically produced titration curve results. It did not include thermal stratification, sediment processes, or other chemical processes. The investigators observed that consideration of these processes did not impair accurate model predictions. Sverdrup and Warfvinge (1988b) reported successful testing on lakes with retention times longer than 0.3 year. Predictions for lakes with shorter retention times were not accurate because a simple one-tank hydrologic model was used.

DeAcid Model

The DeAcid model previously described for dose calculation has the ability to predict reacidification rates. It incorporates the

Table 2.12 Comparison of Reacidification Models Available for Lakes

Model	Reference	Principle	Includes Cost Optimization	Advantages	Disadvantages
Swedish nomograph	Swedish Environmental Protection Board, 1988	Lake retention time and water sampling	No	Simple	Approximation based on empirical data
Sverdrup lake model	Sverdrup, 1985	Mass balance, dissolution model, titration curve	Yes	Simple, field tested, requires few data inputs	One-tank model best for lakes with longer retention times
Acid Lake Reacidification Model (ALaRM)	DePinto et al., 1987, 1989	Mass balance, dissolution model, sediment model, aquatic chemical speciation	Yes	Field tested, comprehensive	Requires considerable data input
DeAcid model	Fraser et al., 1985	Mass balance, dissolution model, titration curve, sediment model	Yes	Field tested, requires few data inputs, computer-assisted program available	None

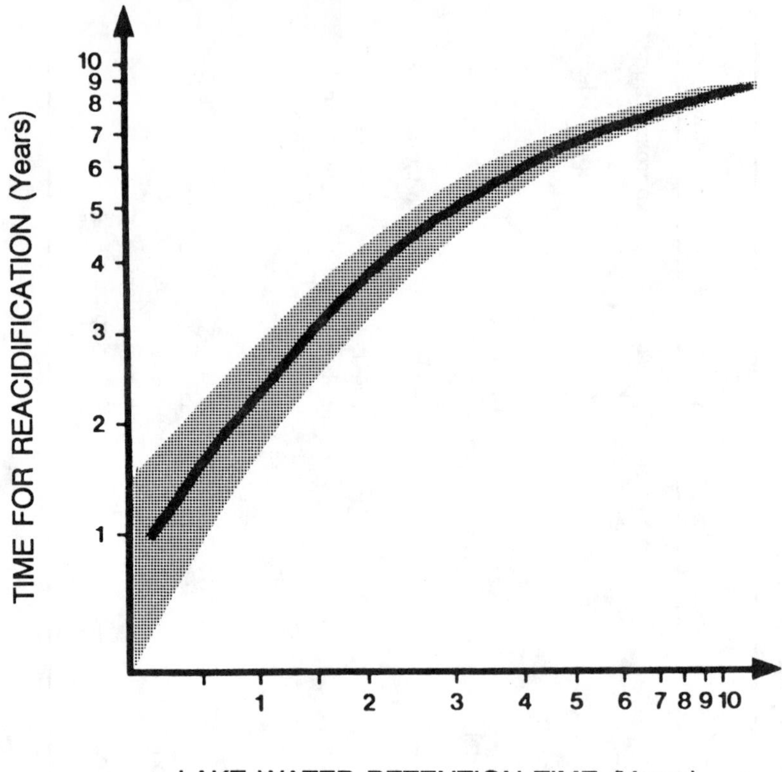

Figure 2.37 Estimated duration of neutralized conditions in lakes after liming. Curve was adapted after Sverdrup et al. (1985, 1986) and is based on empirical information from the Swedish operational liming program. (Source: Swedish Environmental Protection Board, 1988.)

Sverdrup model. Comparisons of model results with actual reacidification of treated lakes have not been reported.

Acid Lake Reacidification Model (ALaRM)

The most comprehensive of lake reacidification models developed to date is a mathematical model described by Scheffe et al. (1988a) and DePinto et al. (1987, 1989). The model is based on water column and sediment mass balances of alkalinity and dissolved inorganic carbon. Unlike the models previously described, this one includes thermal stratification and mixing characteristics in the lake, aluminum chemistry, organic acids, and a sediment submodel such as the one published by Sverdrup and Warfvinge (1985).

Integrated Lake-Watershed Acidification Study Model

This model and other integrated watershed models may be able to predict reacidification rates after liming treatments. Application and refinement of these models to this task, however, have not yet been conducted (Davis, 1988).

Costs of Liming Techniques

Few studies have comprehensively evaluated the costs of liming techniques. Costs cited typically include costs of materials, transportation, and application. The choice of liming material and application technique will affect both the resource costs and requirements. Included here are current costs of selected liming materials, including transportation; information on the historical costs associated with the various liming techniques described in previous sections for lakes, running waters, and watersheds; and case studies on specific analyses of aquatic liming costs. The analyses indicate the wide variability in costs for different situations, such as the cost of available materials, application techniques, and site accessibility. All costs are converted to 1989 U.S. dollars.

Although it may be helpful to know the start-up costs for a particular treatment technique, the actual cost must be amortized over the duration of the treatment. Also, there are other tasks not normally considered in costing an aquatic liming program, including the costs of site assessment, permitting, characterization, monitoring, and administration. These costs can be an order of magnitude higher than material and application costs.

Cost data presented here were often obtained from research and demonstration liming programs. Most of these studies were not conducted with cost optimization as a major consideration. For example, liming materials were often applied much more carefully because of the desire for more precise data on application quantities.

Other complications in comparing costs are that candidate sites for liming are sometimes far from the source of the base material. For example, no limestone quarry is likely to be close to a lake in an area of granite bedrock. This is not a universal problem, however. Unlike parts of Sweden and Norway, limestone vendors in the United States are normally within 50–80 km of candidate sites. Another complication is the availability of labor near the typical acidic lake, which is often remote and inaccessible.

Costs of Neutralizing Agents

The unit costs of the various neutralizing agents may differ from their cost per unit by weight. The costs of the more commonly used alkaline materials (limestone gravel, agricultural limestone, food-grade powdered limestone, hydrated lime, calcined lime, sodium carbonate, sodium bicarbonate, and sodium hydroxide) are shown in Table 2.13. A major cost of governing the total price of the materials is the transportation charge, which is primarily a function of the distance of the material source to the liming site. The material cost in Table 2.13 assumes a transport distance of 80 km within the northeastern United States. Including transportation increases the average cost of the neutralizing materials $28 per tonne.

The costs for neutralizing chemicals vary widely. The tabulation is somewhat misleading because it does not incorporate neutralizing efficiency of the chemical or the variations in the efficiency of application.

Limestone aggregate has the lowest delivered cost per tonne among the selected materials. This material is, of course, used only in limited stream treatment applications, such as rotary drums and limestone barriers. The application efficiency is generally lower for this material, particularly for applications like barriers where the material is not mechanically ground. Limestone powders with a mean diameter of 7.5–14 μm are the most commonly used neutralizing materials in both lake and stream applications. Delivered costs range from $74 to $90 per tonne—much less than the other materials. Other unit costs (such as cost per year or per unit of lake surface area) may be more important values to consider because these may provide results different from those developed in a comparison of material cost per unit of weight. For example, the finer limestone powders (mean diameter 0.7–5.5 μm) cost much more than coarse-grained powders (mean diameter 7.5–18 μm), but because of their greater reactivity the quantity required may be reduced. This relation would also

Table 2.13 Typical Costs, Including Transportation, for Selected Neutralizing Materials

Material	Commercial Classification (mm)	Mean Diameter [a]	Cost [b] ($/tonne)
Limestone aggregate	20–50	38 mm	35
Limestone powder (curve 5)[c]	0–1	18 μm	68
Limestone powder (curve 6)[c]	0–0.5	14 μm	74
Limestone powder (curve 7)[c]	0–0.2	12 μm	80
Limestone powder (curve 8)[c]	0–0.044	7.5 μm	90
Limestone slurry (curve 9)[c]	0–0.005	5.5 μm	135
Limestone slurry (curve 10)[c]	0–0.002	0.7 μm	145
Hydrated lime	Bulk powder		145
Calcined lime	Bulk powder		135
Sodium carbonate	Bulk powder		300
Sodium bicarbonate	Bulk powder		300
Sodium hydroxide	50% solution		400

[a] Surface area-weighted mean.
[b] Assumes transport distance of 80 km within northeastern United States. Does not include provision to download material at site with a fork lift or other device.
[c] Refer to Figure 2.6 for particle size distribution curves.

hold for the lime and sodium-based products, compared to limestones, because of increased dissolution efficiencies. Other factors must be considered, however, such as the possibility of overtreatment and availability.

Limestone slurries are sometimes used in automated dosers and helicopters. The high costs relative to those of powders may be offset in part by the elimination of the need for the labor and equipment involved in a slurrying operation.

Lakes

Costs associated with the various liming techniques for lakes (described earlier) are shown here. The costs are converted to 1989 U.S. dollars by assuming 5% per year inflation. Where possible, costs are presented for the case studies used to illustrate each of the application techniques and materials described earlier. The summary of the cost information presented for lakes shown in Table 2.14 includes costs for materials, transportation, and application; they should be considered as examples only and do not necessarily represent typical costs for each method. Costs may vary because of accessibility, use of donated labor, and cost variations in different locations, including changes in foreign currency exchange rates. Costs of site selection, pre- and posttreatment monitoring, permitting, report preparation, and administration are not included because of the wide variations in these costs among liming projects.

Boat

Costs are presented for manual application of dry powdered limestone, application of limestone by using a slurry box, application of slurried limestone from a barge and pressure tank, and injection of sodium carbonate into lake sediment.

Manual Dry Spreading. Fraser et al. (1984) reported costs for manual spreading of limestone materials from a boat for 30 lakes in Massachusetts. The average of actual costs for treating lakes with a mean surface area of 20 ha were $1800 for limestone (including transportation) and $500 for labor (total cost $2300). This translates to an average cost of about $96/tonne ($115/ha).

Slurry Box Spreading. Costs for the slurry box spreading technique were reported by White et al. (1984) for treating Sandy Lake (74 ha) in Nova Scotia with limestone powder. Based on an average delivered cost of the limestone, the cost for base material was $9900. Another cost reported was $1800 for equipment. Although no application costs were reported, the application process required a field crew of four for 30 days. Judging from this information, the estimated costs were $11,700. The unit cost for treating Sandy Lake was about $87/tonne ($158/ha).

Barge and Pressure Tank. Costs for treating Lake Masen in Sweden by a commercial barge and pressure tank method were reported by Fraser et al. (1985). The 420-ha lake was treated with 1700 tonnes of limestone at a cost of $36,900 for materials and transportation and $24,600 for application. The unit cost for treatment was $35/tonne ($147/ha). The unusually low cost may be due to problems related to foreign currency conversions.

Table 2.14 Historical Costs of Liming Techniques for Lakes[a]

Application Technique	Surface Area (ha)	Cost			Cost/tonne (Cost/ha in Parentheses)			Reference
		Materials[b]	Application	Total	Materials[b]	Application	Total	
Boat —manual dry spreading	20	$1,800	$500	$2,300	$75($90)	$21($25)	$96($115)	Fraser et al. (1984)
Boat —slurry box spreading	74	11,700	—	11,700	87(158)	—	87(158)	White et al. (1984)
Boat —barge and pressure tank	420	36,900	24,600	61,500	22(88)	14(59)	36(147)	Fraser et al. (1985)
Boat —sediment injection[c]	14.7	12,000	20,000	32,000	300(816)	500(1,360)	800(2,176)	Lindmark (1985)
Truck or tractor	114	—	62,000	62,000	—	620(540)	620(540)	Fraser et al. (1984)
Helicopter	25	3,400	22,100	25,500	106(136)	691(884)	797(1,020)	Fraser et al. (1985)
Fixed-wing aircraft	290	28,900	55,700	84,600	170(100)	328(192)	498(292)	Fraser et al. (1984)

[a] Costs are based on actual liming applications after 1970 and are converted to 1989 U.S. dollars assuming 5% per year inflation. Costs vary because of accessibility, use of donated labor, and cost variations in different locations including changes in foreign currency exchange rates. Costs of site selection, pre- and posttreatment monitoring, permitting, report preparation, and administration are not included.

[b] Includes material transportation costs.

[c] Costs are estimated; no actual costs were provided.

Sediment Injection. A review of published literature revealed little information on the costs of injecting sodium carbonate into lake sediments. This author estimated costs for sediment treatment of Lake Lilla Galtsjon on the basis of information provided by Stoclet (1979), Lindmark (1982, 1985), and chemical cost data for sodium carbonate. The 14.7-ha lake was treated with 40 tonnes of sodium carbonate applied in a 10% lake-water solution at an estimated cost of $12,000 for materials and transportation and $20,000 for application. The unit cost was estimated at $800/tonne ($2200/ha). A unit cost comparable to limestone application may be considerably lower if it is assumed that neutral conditions in the lake remain five to seven times longer (Ripl, 1980).

Truck and Tractor

Fraser et al. (1984) provided cost figures for liming Lake Hovvatn, Norway, using a tractor and snowblower on the ice-covered lake and along the shoreline. Total cost for materials, transportation, and application were $62,000, a unit cost of $620/tonne ($540/ha).

Helicopter

Fraser et al. (1985) reported costs for helicopter application of limestone to Woods Lake, New York. The costs for materials and transportation were $3400 for 32 tonnes of limestone slurry. Application costs were $22,100, for a total unit cost of $797/tonne ($1020/ha).

Fixed-Wing Aircraft

Costs for treatment of Trout Lake, Ontario by fixed-wing aircraft were presented by Fraser et al. (1985). The 290-ha lake was treated with 170 tonnes of limestone slurry, costing a total of $28,900 for materials and transport and $55,700 for application—or a unit cost of $498/tonne ($292/ha). The cost per hectare is low compared to that for some other lakes included in Table 2.14, because the limestone dose used to treat this lake was considerably lower.

Streams

The historical costs associated with the various liming techniques for running waters described earlier were not reviewed. As before, the costs are converted to 1989 U.S. dollars. A major portion of the total cost for stream treatment is the initial purchase and installation cost of the treatment device. Other costs include the annual cost of maintenance and labor and the cost of the neutralizing material. Material costs for stream treatment are not included in this discussion, but can be estimated on the basis of costs shown in Table 2.14, and the dosage rate of the particular device.

Where possible, costs are presented for the equipment described in the case studies presented previously. A summary of the cost information presented for streams is shown in Table 2.15.

Dosers

Cost information is presented for electrically powered, battery-powered, and water-powered dosers.

Electrically Powered Doser. Fraser et al. (1985) estimated the consruction costs for a line-powered wet slurry doser to range from $40,000–50,000, based on systems installed in Sweden with an annual dose capacity of 2000 tonnes. Annual labor and maintenance costs for cleaning, lubrication, and general repair were estimated to be about $10,000.

Battery Powered Doser. Consruction costs for a battery-powered doser capable of dosing up to 90 tonnes of dry powdered limestone annually were estimated to be between $7500 and $15,000 (Fraser et al., 1985). Annual labor and maintenance costs were estimated to range from $5000–10,000, depending on the amount of limestone used and maintenance factors, such as frequency of repair and battery replacement.

Water Powered Doser. Fraser et al. (1985) presented costs for the Hallefors water-powered doser then available in Sweden (no longer sold) that is capable of dispensing up to 90 tonnes of dry powdered limestone annually. Construction costs were estimated

Table 2.15 Historical Costs of Liming Techniques for Streams[a]

Application Technique	Annual Dosage (tonnes)	Construction Costs			Annual Maintenance Costs[b]			Reference
		Low	High	Avg	Low	High	Avg	
Doser—electrically powered	2,000	$40,000	$50,000	$45,000	$8,000	$12,000	$10,000	Fraser et al. (1985)
Doser—battery powered	90	7,500	15,000	11,250	5,000	10,000	7,500	Fraser et al. (1985)
Doser—water powered	90	6,000	13,000	9,500	2,500	10,000	6,250	Fraser et al. (1985)
Diversion well	80	16,000	25,000	21,000	5,000	10,000	7,500	Fraser et al. (1985)
Limestone rotary drum	450	99,000	121,000	110,000	12,000	16,000	14,000	Fraser et al. (1984)

[a] Costs are based on actual liming applications after 1970 and are converted to 1989 U.S. dollars assuming 5% per year inflation.
[b] Does not include costs for purchase and transport of neutralizing material.

to be $6000–13,000 and annual maintenance and labor costs and annual maintenance and labors $2500–10,000, depending on the amount of limestone used.

Diversion Well

The estimated current construction cost of a system of two diversion wells capable of delivering 80 tonnes of limestone gravel is $21,000 (Fraser et al., 1985). Annual maintenance and labor costs may range from $5000–10,000.

Limestone Rotary Drum

Current construction costs for a single self-feeding rotary drum installation capable of using 450 tonnes of limestone aggregate annually are estimated to be $110,000 (Fraser et al., 1984). The relatively high cost may be partly offset by the considerably lower cost for limestone aggregate relative to limestone powder or slurry. Annual maintenance and labor costs are estimated at $14,000.

Watersheds

Warfvinge and Sverdrup (1988a) estimated that the annual cost for watershed liming operations is five times that of direct lake application. These estimates assume a lake retention time of 1 year and reacidification times for lake and watershed liming of 3 and 6 years, respectively. Available information suggests that the reacidification time estimated for watershed liming is low. If so, the economics of each technique may be more similar if considered on an annual basis.

Case Studies

Only a few studies of liming costs and benefits have been published (Menz and Driscoll, 1983; Riely and Rockland, 1988; R.W. Brocksen, personal communication, 1990).

Adirondack Lakes

Menz and Driscoll (1983) computed the costs for liming 663

lakes in the Adirondack region of New York. Total costs included the cost of the limestone material to meet a target ANC level and the costs of application. Separate costs were given for accessible and remote sites. Assuming the application of a 5-year dosage, estimated costs averaged $53–107/ha for accessible lakes and $637–947/ha for remote lakes (converted from 1982 to 1989 U.S. dollars). Using the data base of Menz and Driscoll (1983), Dutkowsky and Menz (1985) presented a cost function for mitigation by liming. The results were reported to provide an accurate model for predicting lake neutralization costs, particularly for the Adirondack region.

Living Lakes, Inc.

R.W. Brocksen (personal communication, 1990) provided typical information on the costs of material, transportation, and application based on experience with treatment of 35 sites under the LLI program. Costs for fine grades of bulk limestone (average particle diameter, 6–20 μm) ranged from $40–80/tonne. Bagged material typically costs about 10% more. If the material is slurried before distribution by barge or helicopter, costs may increase by 30%.

Transportation of material varied from $20–50/tonne, depending on the quantity being transported. Transporting more material would reduce the cost per tonne. Also, a premium is charged for transporting less than full truckloads. These transportation costs do not include the costs of unloading bagged material at the site (normally $100–200).

Labor costs for application of limestone ranged from $400–800 to treat a lake of 4–30 ha when a boat and slurry box were used. This translates to unit costs ranging from $95–130/tonne. Total labor costs for application by a barge and pressure tank technique for accessible lakes lager than 50 ha ranged from $130–150/tonne.

Cost/Benefit Analysis

Riely and Rockland (1988) presented a method for evaluating the benefits associated with liming acidic waters and how these benefits compare with the cost of liming. The benefit-cost analysis was seen as a useful tool to allow the decisionmaker to understand whether a particular liming project is justified from an economic perspective.

OTHER TECHNIQUES FOR MITIGATION

This section describes techniques other than liming that have been tested or proposed for mitigation of acidic surface waters; these include nutrient addition, stocking of acid-tolerant or -resistant fish strains, and pumping of alkaline water. The advantages and disadvantages of these methods are summarized in Table 2.16. Little information is available for evaluating the ability of these methods to mitigate the adverse ecological effects of acidic surface waters.

Nutrient Addition

Two methods have been suggested to mitigate ecological effects of acidic surface waters—the addition of phosphorus and the addition of an organic carbon source. No studies have thoroughly evaluated the value of these methods for mitigation. The possible use of these methods, based on other related studies of lake ecology, are explored here.

Phosphorus Addition

Most acidified surface waters are generally thought to be phosphorus limited for algal growth. It has been postulated that fertilization of acidic lakes by phosphorus addition may assist in reducing acidic conditions. A second benefit to acidic lakes may be a replenishment of the biomass and increase in fish food supply.

Use of this method implies that uptake of nutrients during primary product will result in the consumption of protons and the corresponding generation of alkalinity (Altshuller and Linthurst, 1984). The concept of alkalinity changes generated by phytoplankton growth has been studied by Brewer and Goldman (1976). Simplistically, the method suggests that nutrients are taken up by algae during photosynthesis and result in the consumption of protons. This consumption generates alkalinity by the assimilation of nitrate and a corresponding uptake of hydro-

Table 2.16 Advantages and Disadvantages of Mitigation Techniques Other than Liming

Technique	Relative Use	Advantages	Disadvantages
Phosphorus addition to lakes	Uncommon	Stimulates phytoplankton growth in unproductive lakes. May increase proton consumption (generating alkalinity).	Uncertain and unproven technique. Added phosphorus may precipitate due to high aluminum concentrations. Public opposition.
Organic carbon addition to lakes	Uncommon	Reduces toxicity of metal ions through complexation. May buffer against further acidification.	Uncertain and unproven technique. Added organic material may cause oxygen depletion and lead to anoxic conditions. Public opposition.
Stocking acid-tolerant fish strains in lakes and streams	Occasional	Increased survival of stocked fish species. May avoid the need to neutralize mildly acidic waters.	Some field tests have not shown significant difference between acid tolerant and normal fish stocks. Even acid tolerant strains cannot survive in highly acidic waters (pH ≤ 5).
Pumping alkaline water to lakes and streams	Uncommon	Avoids the need to add base materials. Chemical effectiveness fairly certain.	Possible depletion of groundwater reserves. Applicability may not be widespread. Requires nearby source of electricity. May require continuous pumping.

gen ions (Turner et al., 1990). Unfortunately, there are a number of competing factors. For instance, an uptake of ammonia or organic nitrogen could occur, resulting in the consumption of alkalinity (Turner et al., 1990). Also, the reduction of nitrate—a process that may occur in sediments—would produce alkalinity.

The method of phosphorus addition has not been adequately tested in the field for acidic lakes. Usually, phosphorus is added in combination with a neutralizing agent such as limestone (Dillon et al., 1979; Hultberg and Andersson, 1982). Phosphorus has also been added to acidic lakes through the addition of wastewater effluent, but these experiments were not designed to evaluate the ability of the phosphorus alone to generate additional alkalinity (Scheider et al., 1975; Hultman et al., 1983).

Aluminum sometimes interferes with the availability of phosphorus by biota. When phosphorus is added to a lake, there is competition between phytoplankton and aluminum ions for the phosphorus molecule. Phosphorus is often coprecipitated with aluminum in acidic lakes (Driscoll et al., 1982, 1988b). Consequently, phosphorus has usually been added after liming, when water column concentrations of aluminum are lowered (Dillon et al., 1979).

Because most of the experimental studies of phosphorus addition to acidic lakes have been performed in combination with a neutralizing agent or wastewater, it is not possible to determine whether phosphorus additions alone resulted in the generation of alkalinity. The use of phosphorus in combination with other methods may be beneficial to biota in acidic water bodies, but it is unlikely that this method alone is efficient in generating alkalinity. Furthermore, there may be opposition to the addition of fertilizers to lakes.

Addition of Organic Carbon

Acidified lakes are often low in dissolved organic carbon, a condition that affects the biological health of a lake by causing a loss of alkalinity and buffering capacity. The resulting water body may be more susceptible to further acidic inputs. Low dissolved organic carbon also minimizes the possibility of organic complexation of metal ions. Organic acids can complex with potentially toxic metal ions such as aluminum and mitigate the toxicity of the metal to aquatic organisms (Driscoll et al., 1982, 1989b). Degerman (1987), however, found that metals bound to organic acids can be liberated at certain pH values.

The addition of organic carbon sources may help buffer a water body from further acidic inputs and may reduce metal toxicity. Although this would not neutralize an already acidic water body, it might improve conditions for fish survival. Possible carbon sources are commercial chemicals such as citrate and oxalate, treated wastewater, and leaf leachate.

Few experiments have been reported in the literature on the concept of the addition of organic carbon to acidic waters. Hultman et al. (1983) used treated domestic wastewater to improve the biological productivity of an acidic lake. The wastewater was used primarily as a phosphorus source; consequently the impact of organic carbon addition on water quality and biota was not evaluated. Scheider et al. (1975) added treated wastewater to enclosures in a limited lake. This experiment was also designed to evaluate the effectiveness of wastewater as a phosphorus source.

An undesirable effect of organic carbon addition may be the development of anoxic conditions due to the oxygen demand exerted by organic materials.

The lack of available data makes it extremely difficult to evaluate the potential for organic carbon additions alone to improve conditions in acidic lakes. The cost of commercial carbon sources may be prohibitive. As in phosphorus addition, there may be considerable justifiable opposition to the addition of a more cost-effective organic carbon source such as wastewater to acidified but otherwise pristine lakes.

Stocking Acid-Tolerant or -Resistant Fish Species

It is well known that some fish species are more tolerant than others of acidic conditions in lentic and lotic waters (Gjedrem, 1980). Stocking more resistant fish species, however, may not provide the desired fishery for an acidic lake or stream.

It has also been shown that there are differences in acid tolerance between different strains of fish of the same species (Schofield et al., 1981). It may thus be possible to enhance fish survival, and sometimes growth and reproduction, by stocking acid-tolerant or acid-resistant strains of desired fish species in surface waters where water quality conditions are not completely limiting to fish stocks. It should also be noted that young "acid-resistant" fish species may be more sensitive than adults, potentially limiting reproduction. Food availability is also a potential problem.

The objective of this technique is solely to improve the survival of stocked fish species. There is no improvement in water quality or in the status of other aquatic biota.

Acid-tolerant strains of fish have been stocked in the Adirondack region of New York (Simonin et al., 1988). In waters being managed by the New York Department of Environmental Conservation, stocking resistant fish strains was not seen as a viable alternative to the existing liming program. Acid-resistant strains of brook trout (the primary stocked species) were not considered to be able to survive in most acidic waters managed by the department. For example, Simonin et al. (1988) reported that "acid-tolerant" strains of brook trout may be able to survive in a lake with a pH as low as 5.1 compared to pH 5.5 for normal acid-sensitive strains.

Stocking strains more tolerant of acidic conditions may be useful in conjunction with liming to protect against storm episodes. These episodes may result in temporary decreases in pH to levels where only acid-tolerant fish strains can survive.

Pumping of Alkaline Water

Pumping of water from a nearby source that contains alkalinity has been suggested as a viable technique for the neutralization of acidic surface waters (Gagen et al., 1989; D. Knauer, personal communication, 1990). For example, it may be possible to pump water from a deep groundwater source, which would normally be more alkaline than nearby surface waters. Another possibility would be the pumping of water from another surface water source of higher alkalinity.

Wisconsin Studies

The Wisconsin Department of Natural Resources investigated the water requirements for neutralizing seepage lakes by pumping alkaline groundwater (D. Knauer, personal communication,

Table 2.17 Range in Aluminum Concentrations and pH of Water in a Pennsylvania Stream Neutralized by Additions of Groundwater. (Source: Fisher, 1985.)

Stream Section	Aluminum (μg/L)	pH
Untreated	282–588	4.62–4.98
Treated	7–112	5.39–6.78
Nonacidic control	5–24	6.33–6.91

1990). The integrated Lake-Watershed Acidification Study model was used to provide the hydrologic and chemical equilibrium routines required for the analysis. It was determined that the process was technically feasible, but it has not yet been applied to any lakes in Wisconsin.

Pennsylvania Studies

The pumping of alkaline groundwater has been reported in Pennsylvania for the temporary neutralization of an acidic stream (Fisher, 1984, 1985; Gagen et al., 1989). Groundwater was pumped from wells to neutralize an acidic section of Linn Run. The researchers were interested in determining if the method would allow the section to sustain a put-and-take trout fishery.

Three sections of the stream were evaluated. One section was treated and a second section immediately upstream was monitored as the untreated control. Water quality and catch rates were also determined for a downstream section of the stream in which pH was in the circumneutral range.

Three wells were used to pump alkaline water to the treatment section of the stream. Concentrations of aluminum and H^+ were lowest in the treated section of the stream (Table 2.17). Catch rates of stocked brown trout and brook trout were similar for the nonacidic and treated sections of the stream when alkaline groundwater was pumped to maintain the pH of the treated section above 6.0. No fish were in the acidic untreated section of the stream.

An important consideration noted by the researchers was the possible depletion of groundwater reserves by continuous pumping. The method was considered to successfully neutralize the stream section to establish a short-term put-and-take fishery and to provide recreational value to the stream. It is not known whether the method has wide applications or whether the costs of treatment compare favorably to those of other mitigation options such as liming.

3

Effects of Liming on Water Chemistry and Biota

"The best never let a little rain stand in their way."
Gene Kelly

INTRODUCTION

The addition of base materials to surface waters is designed to perturb aquatic systems and thus to improve water quality and to enhance the production of favored biota. Effects have been shown to be different for lakes compared to streams and different for various types of aquatic systems. The dynamic nature of the neutralization process results in both short- and long-term effects to the aquatic environment. Also, major changes occur in water quality and biota as water bodies reacidify.

In this section will be reviewed the physical, chemical, and biological effects of liming lakes, streams, and their watersheds. The results of monitoring the effects of liming for the major mitigation programs in the United States are summarized, physical and chemical effects of liming are described in general, and biological effects are also discussed.

RESULTS OF MAJOR MITIGATION PROGRAMS

This summary of recent liming experiments conducted by the major mitigation programs in the United States is included to provide a comprehensive listing of the major mitigation programs and the specific aspects of physical, chemical, and biological effects studied. It serves as an introduction to the general evaluation of the effects of liming covered later.

Results reported for the major mitigation programs in the United States are described in more detail in Appendix C. Because much of the important literature on effects of liming was obtained from studies outside the United States, experiments conducted by mitigation programs in Canada, Sweden, Norway, and Scotland are summarized in Appendix D. Unlike the more comprehensive descriptions of all major experimental mitigation programs in the United States (Appendix C), the results described in Appendix D only highlight one or two important projects in each country.

Table 3.1 summarizes the major experimental liming projects conducted in the United States and includes the lakes or streams treated, the particular aspects studied, and appropriate references. Only a few of the surface waters that have been limed in the United States were later evaluated in sufficient detail to provide information on the chemical and biological effects of the liming. The projects were sponsored by two not-for-profit organizations (Electric Power Research Institute and Living Lakes, Inc.), the U.S. Fish and Wildlife Service, and two state agencies (Maryland and West Virginia). Many of the studies were jointly funded by the organizations.

Most of the programs involved lake liming; only recently have a number of new projects been started on streams. No information is available on the effects of watershed liming treatments; however, a study of liming the Woods Lake watershed in New York was begun in 1989 with the cooperative support of Electric Power Research Institute, Living Lakes, Inc., the U.S. Fish and Wildlife Service, and the Empire State Electric Energy Research Corporation.

The Electric Power Research Institute is conducting an aquatic liming study in the Adirondack region of New York State, termed the Lake Acidification Mitigation Study. Three lakes—Cranberry Pond, Woods Lake, and Little Simon Pond—were treated. A comprehensive evaluation of physical, chemical, and biological effects was reported in several publications.

Liming effects on 10 small fishless lakes in New York were studied under the direction of the U.S. Fish and Wildlife Service in cooperation with Cornell University. Chemical and biological data collected by the New York Department of Environmental Conservation and by private organizations were used in the evaluations.

The U.S. Fish and Wildlife Service also has a major mitigation research program in place to evaluate experimental liming of one lake in Minnesota and three streams—one each in Massachusetts, West Virginia, and Tennessee. These investigations are focused on comprehensive evaluations of water quality and fish.

The water quality and biological results of automated doser stream liming was evaluated for two streams in Maryland. This study provided information on the effectiveness of stream dosing during storms to restore favored fish species.

An analysis of pre- and postliming water quality for 22 lakes was the focus of a study by Marcus (1988) of liming conducted as part of the Living Lakes program during 1986 and 1987. Additional information was provided by Brocksen and Emler (1988) on liming effects on 10 streams in the northeastern United States.

EFFECTS ON WATER CHEMISTRY

The addition of base materials to surface waters may cause physical and chemical changes to aquatic ecosystems. In this chapter, the results of the liming experiments in the United States, Canada, Sweden, Norway, and the United Kingdom in which the effects on water chemistry were evaluated are integrated and summarized.

This discussion primarily reflects the findings of studies reporting the effects of liming directly to lakes. Considerably less information exists on the effects of direct addition to streams and watersheds. The focus is also on results of studies evaluating the effects of limestone addition, rather than other neutralizing agents, although the uses of other materials are described when results are unique to that material.

Physicochemical Effects on Lake Water Column

Changes in transparency, color, and thermal conditions are among the physical effects that may occur after the addition of base materials to surface waters. Chemical changes that normally occur after liming include increases in pH, ANC, dissolved inorganic carbon, and calcium concentration. Other possible changes include increases or decreases in organic matter, nutrients (phosphorus and nitrogen), other major anions and cations, and trace constituents (such as aluminum, manganese, and zinc). The

Table 3.1 Summary of Major Experimental Mitigation Projects in the United States

Type of Water or Watershed and Program	Lake or Stream	Aspects Studied	Reference
Lakes			
Electric Power Research Institute	Three lakes in Adirondack Park, NY	Background information	Porcella, 1989
		Morphometry	Staubitz and Zarriello, 1989
		Dissolution efficiency	Driscoll et al., 1989
		Water chemistry	Fordham and Driscoll, 1989
		Calcium in lake sediments	Young et al., 1986, 1989
		Phytoplankton	Bukaveckas, 1988a, 1989
		Sediment algae	Roberts and Boylen, 1989
		Zooplankton	Schaffner, 1989
		Aquatic insects	Evans, 1989
		Stocked fish	Gloss et al., 1989b
		Reacidification	Schofield et al., 1989
U.S. Fish and Wildlife Service	Ten small fishless lakes in the Adirondacks	Morphometry	Schofield et al., 1986; Gloss et al., 1989a
		Water chemistry	Schofield et al., 1986; Gloss et al., 1989a
		Fisheries	Schofield et al., 1986; Gloss et al., 1989a
		Reacidification	Gloss et al., 1988a,b
		Management guidelines	Gloss et al., 1989a
U.S. Fish and Wildlife Service	Thrush Lake, Minnesota	Background information	Schreiber, 1988; Schreiber et al., 1988
		Water chemistry and fish	R.K. Schreiber, personal communication, 1990

Living Lakes, Inc.	Twenty-two lakes in the northeast U.S.	Background information Water chemistry	Brocksen and Emler, 1988 Marcus, 1988
Streams			
U.S. Fish and Wildlife Service	Three streams: Mass., W. Virginia, Tenn.	Background information Water chemistry and fish	Schreiber, 1988; Schreiber et al., 1988 R.K. Schreiber, personal communication, 1990; Brocksen and Emler, 1988
Living Lakes, Inc.	Ten streams in the northeast U.S.	Background information Water chemistry	Adams and Brocksen, 1988
Maryland Department of Natural Resources	Two streams in Maryland	Water quality and biota	Janicki and Greening, 1988a,b
Watersheds			
Cooperative effort (EPRI, LLI, FWS, and ESEERCO)[a]	Woods Lake	Background information Effects	Brocksen et al., 1988 (Not yet available)

[a] Electric Power Research Institute, Living Lakes, Inc., U.S. Fish and Wildlife Service, and the Empire State Electric Energy Research Corporation.

Table 3.2 Physicochemical Effects of Liming Surface Waters

Variable	Typical Response; increase (+), Decrease (-), or Similar (o)		Comments
	Clear Waters [a]	Colored Waters [b]	
Physical Effects			
Transparency	(-)	(+)	Short-term response is sometimes a decrease in colored waters until dissolution or settling of the limestone
Color	(+)	(-)	Often the opposite response to transparency
Temperature	(+)	(-)	Often the opposite response to transparency
Chemical Effects			
pH	(+)	(+)	
Acid neutralizing capacity	(+)	(+)	
Dissolved inorganic carbon	(+)	(+)	
Base cations	(o)	(o)	Except for cations associated with the base material
Acidic anions	(o)	(o)	Except for anions associated with the base material
Phosphorus	(+)	(+)	Increase is often smaller than instrument detection limit
Nitrogen	(o)	(o)	Organic nitrogen and nitrate have been reported to increase slightly in some studies

Dissolved organic carbon	(+)	(-)	
Toxic metals	(-)	(-)	Particularly aluminum (see Table 3.3)

[a] Clear waters generally are those with color less than 50 units and dissolved organic carbon less than 5 mg/L.
[b] Colored waters are humic systems with color generally greater than 50 units and dissolved organic carbon greater than 5 mg/L.

physical and chemical changes that usually occur after limestone neutralization of surface waters are summarized in Table 3.2.

There are often distinct short-term changes (within days or weeks) that are sometimes very different from changes over longer terms (months). Both short-term and long-term changes in the physical and chemical characteristics of surface waters after liming are described here.

There have been certain consistencies in the changes that occur after liming, and several examples are cited. Other characteristics of surface waters have been reported to increase, decrease, or remain the same after liming.

Changes are often immediate after direct addition of base materials to surface waters, but are normally much slower for watershed liming. Changes in lake and stream water usually do not occur until precipitation reacts with base material deposited on forest soils or wetland areas (Rosseland and Hindar, 1988; Brown et al., 1988).

Physical Effects

Physical changes that may occur after the addition of base materials to surface waters typically include increases or decreases in transparency, color, and thermal conditions. These changes are often interrelated.

Transparency

Many researchers have reported immediate increases in transparency after liming to lakes (Dickson, 1979; Yan and Dillon, 1984; Hultberg and Andersson, 1982; Driscoll et al., 1982). This increase normally occurs only in humic, colored lakes, where liming can cause organic colloids to precipitate from solution (Gloss et al., 1989a). Yan and Dillon (1984) reported that this response occurred after the liming of Sudbury lakes and was due to the removal of dissolved organic matter by coprecipitation with metals.

Roberts and Boylen (1989), however, reported a decrease in transparency for several weeks after liming Woods Lake in the Adirondacks. The transmittance, measured at a depth of 1 m with a photometer, decreased from 71% to 43% over a 2-week period after liming. Similarly, transparency decreased due to undissolved limestone during a 2-week period after limestone treatment of Thrush Lake in Minnesota (R.K. Schreiber, personal communication, 1990). Gloss et al. (1989a) reported a significant initial decrease in transparency in only 1 of 10 limed lakes in the Adirondack region of New York. Wright (1985) suggested that short-term decreases in transparency may be due to pH-dependent light absorption by humic substances.

Long-term changes in transparency are often due to biological factors. For example, transparency has been found to decreased due to increases in phytoplankton. Increased respiration in the sediment may also release organic substances to the overlying water, resulting in decreased transparency (Wright, 1985). Hultberg and Andersson (1982) attributed an observed decrease in transparency in two of four limed lakes in Sweden to changes in phytoplankton composition.

Color

Another physical change that may occur immediately after liming is a change in color. This change is sometimes directly related to transparency, i.e., an increase in transparency is accompanied by a decrease in color and vice versa (Hasselrot and Hultberg, 1984). There is no consistent response, in part due to natural differences in color of the water in different lakes. Color normally decreases after liming in humic, highly colored lakes because of precipitation of organic materials from solution (Degerman, 1987; Degerman and Nyberg, 1989).

Thermal Changes

Changes in transparency and color of lakes often results in changes in thermal regimes. The absorption of solar energy is directly related to transparency. When transparency decreases after liming, water temperature in the epilimnion may increase. A corresponding decrease in water temperature in the hypolimnion may also occur because the deeper water may be more shaded. The opposite can occur when transparency increases because solar absorption decreases in upper layers and increases in deeper waters.

Yan (1983) indicated that Secchi disk transparency in Lake Lohi, Ontario was positively correlated with thermocline depth, thickness of the epilimnion, and hypolimnetic heating rate. Hultberg and Andersson (1982) concluded that variations in light penetration in the water column caused by liming could result in changes in the heat budget, mixing characteristics, and oxygen profiles of a lake, and could be of significance to biological communities. Few studies have actually observed changes in thermal regimes that could be attributed solely to liming. Other factors, such as differing weather patterns from year to year, make such a comparison difficult.

Chemical Changes

Chemical changes that normally occur after liming include increases in pH, ANC, dissolved inorganic carbon, and calcium concentration. Other possible changes include increases or decreases in organic matter, nutrients (phosphorus and nitrogen species), other major anions and cations, and metals that are potentially toxic to aquatic organisms (such as aluminum, manganese, and zinc).

pH

Initial changes in acid-base chemistry after limestone addition include an increase in pH. Changes in pH can be dramatic immediately after liming. Low initial inorganic carbon often allows immediate increases in pH due to the lack of carbonate buffering. For example, Fordham and Driscoll (1989) observed pH increases in the upper mixed waters from 4.9 to 9.4 in Woods Lake and from 4.6 to 9.1 in Cranberry Pond. The fineness of the limestone that was used caused dissolution to exceed the rate of intrusion of atmospheric CO_2. The pH values decreased over a 4-week period as the CO_2 content in the water equilibrated with the atmosphere.

Immediate increases in pH to values above 9 were also reported in Minnesota (R.K. Schreiber, personal communication, 1990) and New York (Gloss et al., 1989a). In these studies, finely ground limestone was applied to the whole lake. Other whole-lake liming experiments, however, did not result in pH values above 9. Dillon et al. (1979) reported initial pH increases to near 8 after whole-lake liming, and Molot et al. (1986) observed pH increases slightly above 8 immediately after Bowland Lake was limed.

Generally, the application of limestone to ice-covered lakes in winter or to watersheds results in less dramatic and more gradual increases in pH. The changes, of course, depend on the amount applied, the particle size of the material, and climatic and hydrologic factors. According to Wright (1985), limestone applications to the ice-covered Lake Hovvatn and Pond Pollen resulted in increases in pH to maxima of only 6.5 and 7.4, respectively. Brown et al. (1988) observed subwatershed stream pH increases approaching 8 when limestone was applied to subwatersheds of Loch Fleet. The maximum pH recorded at the lake outlet did not exceed 7. Watershed liming at Lake Tjonnstrond resulted in increases in pH from 4.5 to a maximum of 7.1 after the first rain event (Rosseland and Hindar, 1988).

Also, increased pH has been found to be maintained longer after watershed liming than after whole-lake liming (Fraser et al., 1984; Porcella, 1988; Brown et al., 1988; Rosseland and Hindar, 1988).

Acid Neutralizing Capacity, Dissolved Inorganic Carbon, and Major Ions

In addition to increases in pH, limestone application results in corresponding increases in ANC, calcium, and dissolved inorganic carbon. Variations in the chemical response of surface waters to liming are due to the following primary factors:

- chemical characteristics of the water
- thermal and hydrologic characteristics of the water
- particle size and chemical characteristics of the neutralizing material
- method of base application
- amount of base applied
- response of aquatic biota (e.g., sphagnum die-off)

Important chemical factors include the initial pH, ANC, and dissolved inorganic carbon. When these characteristics are higher before neutralization, e.g., positive ANC and pH above 5, there is enough buffering against rapid immediate changes. Fordham and Driscoll (1989) suggested that low initial inorganic carbon in Woods Lake and Cranberry Pond was an important factor in the dramatic immediate increases observed in pH and ANC after liming. The extremely low inorganic carbon in both lakes before neutralization provided little buffering. The lakes first responded like systems closed to atmospheric CO_2.

Thermal changes in the surface water over time profoundly affects longer term acid-base chemistry. For example, in lakes where complete spring and fall overturn occurred after liming, Gloss et al. (1989a) reported increased pH, ANC, and calcium concentrations in upper waters and decreased concentrations in deeper waters after mixing and restratification. Lakes that did not completely overturn maintained high ANC and calcium in all but the surface layer. This was due to the isolation of the deep-water source of alkalinity.

Watershed hydrology is a major factor in the loss of ANC, dissolved inorganic carbon, and calcium over time (Fraser et al., 1984; Porcella, 1988; Gloss et al., 1989a). Simple flushing of neutralized water occurs for lakes with short hydraulic retention times (Driscoll et al., 1989a). Further, the lack of watershed buffering capacity can allow seasonal flows of highly acidic water through lakes. This can result in rapid decreases in pH and depletion of ANC. Spring snowmelt and storm events, for instance, can reintroduce highly acidic waters that also contain lower inorganic carbon and calcium than the neutralized waters. In certain lakes the highly acidic waters can travel along the surface of the lake or rapidly mix throughout the water column depending on thermal conditions (Gloss et al., 1989a). Acidic events can also profoundly influence the chemistry of streams that have neutralization devices (such as limestone barriers and diversion wells) that may not adequately adjust dosages for high flow rates.

Particle size of the neutralizing material is also an important factor. Large limestone particles may settle to the bottom of a lake or stream without dissolving. This could result in inadequate dissolution if not accounted for in the dosage calculations. Very small limestone particles, on the other hand, may dissolve rapidly and increase initial pH excessively.

The very small limestone particles used in the liming of Woods Lake and Cranberry Pond in New York and Thrush Lake in Minnesota caused very rapid dissolution and immediate pH values above 9 (Fordham and Driscoll, 1989; R.K. Schreiber, personal communication, 1990). Also, the settling velocities of these small particles were so low that they were unable to penetrate the thermocline, resulting in confinement of the limestone dose to the upper layers of the water column. An alternate strategy, in which a lower initial dose of very fine limestone is coupled with a dose of larger particles for sediment treatment, may be preferred. This approach may result in more effective neutralization throughout the water column and may possibly delay reacidification through delayed neutralization from limestone particles in the sediment. This approach was used in the retreatment of Woods Lake in 1986, but results have not yet been published (Porcella, 1989).

Base cations and acidic anions not associated with the neutralizing material (e.g., other than calcium from limestone or magnesium from dolomite) do not appear to be altered by neutralization. Marcus (1988), who analyzed major cations and anions from 20 lakes before and after neutralization with high-calcium limestone, found no significant differences between pre- and posttreatment samples for magnesium, sodium, potassium, chloride, and sulfate. Levels of ANC, calcium, dissolved inorganic carbon, and pH, on the other hand, were significantly higher in posttreatment samples.

Many other researchers also reported no significant change in base cations and acidic anions (except those associated with the neutralizing material) after the neutralization of lakes and streams (Hultberg and Andersson, 1982; Wright, 1985; Driscoll et al., 1989). Degerman and Nyberg (1989) analyzed water chemistry before and after liming 112 lakes in Sweden and reported increases in only pH, conductivity, alkalinity, and calcium.

Dudley and Bradt (1987) observed increases in both calcium and magnesium after limestone addition and attributed the increases to the chemical characteristics of the limestone. Lindmark (1985) observed increased sodium levels in Lake Lilla Galtsjon after the injection of sodium bicarbonate in the sediments. Siegfried et al. (1987) also reported an expected increase in sodium levels after the sodium bicarbonate treatment of Bone Pond in the Adirondacks.

Nutrients and Organic Matter

The concentrations of nutrients and organic matter may change in lake and stream water after treatment with base materials. No studies, however, have shown consistent responses of phosphorus, ammonia, nitrate, and dissolved organic carbon in limed lakes.

Phosphorus has sometimes been reported to increase slightly in lake water after treatment with limestone (Fraser et al., 1982; Wright, 1985; Hasselrot et al., 1984; Hornstrom and Ekstrom, 1986; Fraser et al., 1985; Porcella, 1989). This occurrence may be beneficial because phosphorus is the limiting nutrient in many acidic oligotrophic lakes (Almer et al., 1978; Dillon et al., 1979; Broberg and Persson, 1984). The relation between phosphorus and biological productivity in acidic lakes is discussed by Baker et al. (1990a).

The reported increase in phosphorus after liming is probably attributable to either of two possible explanations: (1) trace amounts of phosphorus contained in the limestone or (2) release of phosphorus from lake sediments. Two other explanations for increased phosphorus have been suggested: reduced precipitation of phosphorus by aluminum (Almer et al., 1978) and increased phosphatase activity (Olsson, 1983).

Conditions of increased phosphorus, however, have not been well documented. Most studies indicate no significant change in phosphorus after liming. Reported changes have often been within optimum instrument detection limits. Marcus (1988) analyzed phosphorus levels in 20 lakes before and after neutralization with high-calcium limestone. No significant differences were found between pre- and posttreatment samples. Yan and Dillon (1984) observed no differences in pre- and posttreatment phosphorus for four lakes near Sudbury, Ontario.

Wright (1985) calculated the increase in phosphorus expected on the basis of complete dissolution of phosphorus content in the limestone. Judging from these calculations, an increase of 1–2 μg/L was expected in Lake Hovvatn and 6–12 μg/L in Pollen Pond. These increases were near the values observed. Hultberg and Andersson (1982), who also calculated the expected increase in phosphorus in lake water after treatment, reported a theoretical increase of 10 μg/L. No significant change in phosphorus, however, was observed. Broberg (1988), who found a slight increase in phosphorus after liming that was about equal to the expected phosphorus increase due to trace concentrations in the limestone, could not attribute the increase to the limestone because the normal spring overturn was observed to contribute the same increase in lake phosphorus concentration.

The probability is low that limestone purchased commercially, especially in the contiguous United States, could cause any detectable changes in in-lake phosphorus levels. A dose of 10 mg/L $CaCO_3$ and a maximum phosphorus content of 0.01% would yield a potential total phosphorus change of 1 μg/L.

The release of phosphorus from sediments has been suggested as a likely cause of increased water-column phosphorus after liming, because the liming can increase respiration in sediments. Wright (1985) suggested that this may have occurred in one of the lakes he studied because the increased phosphorus was of long duration, suggesting a continued supply rather than that coming from a one-time addition of limestone. Also, the elevated phosphorus levels roughly paralleled elevated total organic carbon levels. Both organic carbon and phosphorus might be released together when increased respiration in sediments occurs. On the other hand, Wright (1985), in a laboratory evaluation of sediment phosphorus release, found only minor input to the overlying water.

Short-term increases in phosphorus may be related to pH shock, which can provide a source of dead phytoplankton for decomposition. This die-off may result in subsequent (but unsustained) phosphorus release from the sediments.

The effects of liming on nitrogen species is also somewhat contradictory in the literature. Wright (1982, 1985) reported no increase in ammonia or nitrate in either Lake Hovvatn or Pollen after liming, although total nitrogen concentrations increased slightly. Total nitrogen concentrations increased steadily for 1 year in Lake Stora Holmevatten, Sweden and then decreased to a concentration still somewhat higher than that before limestone addition (Hasselrot et al., 1984). Marcus (1988), in a statistical evaluation of 20 lakes, found that total nitrogen and nitrate tended to be greater after liming. These changes, however, were subtle, particularly for nitrate; maximum levels were 1.71 mg/L before liming and 1.90 mg/L after liming.

Broberg (1988), who conducted a thorough evaluation of nitrogen species before and after liming in Lake Gardsjon, Sweden, observed a decrease in organic nitrogen immediately after liming, but the decrease was less than that in the untreated reference lake. The longer term response was an increase in organic nitrogen, reportedly due to mineralization of sphagnum and increased sediment respiration.

Evidently, limestone addition can increase organic nitrogen and possibly nitrate levels, primarily because of changes in the make-up and density of aquatic biota. Some studies, however, indicated no long-term change in total nitrogen, ammonia, or nitrate levels (Baalsrud, 1985).

Liming has often been shown to affect organic carbon concentrations in the water column, but increases have been reported in some studies and decreases in others. The statistical evaluation of 20 limed lakes by Marcus (1988), however, revealed no significant change in dissolved organic carbon.

Liming can decrease organic carbon content in the water column by precipitating humic substances. This phenomenon would normally occur only in lakes with significant humic matter. Decreases in color and increases in transparency normally occur along with decreased organic carbon content. Several lakes in Sweden responded in this way (Hultberg and Andersson, 1982), as did some Adirondack lakes (Driscoll et al., 1982). This precipitation may not be desirable, because organic carbon can be an important pH buffer and its removal may hasten reacidification. Also, organic anions can complex trace metals such as aluminum, resulting in reduced metal toxicity to aquatic organisms (Driscoll et al., 1980; Baker et al., 1990a).

Increases in organic carbon have also been observed after liming. These increases usually coincide with decreases in transparency and increases in color. Wright (1985) found increases in total organic carbon in Lake Hovvatn after liming and suggested that the increases may have been due to a combination of biological factors. For example, increased organic carbon concentrations have been found to be associated with increased phytoplankton growth. Increased respiration in the sediment may also release organic substances to the overlying water (Wright, 1985).

Broberg (1988) reported that photodegradation of organic matter was inhibited by liming; he found higher dissolved organic carbon after the liming of Lake Gardsjon and suggested three possible reasons: inhibition of photodegradation, increased sediment respiration, and mineralization of organic carbon in sphagnum and benthic mats. Also observed was an initial lowering of particulate organic carbon, evidently due to coagulation and settling from the water column. The longer term result was an increase in both dissolved and particulate organic carbon. The increase in particulate carbon was reportedly due to the breakdown of sphagnum and benthic algae. The increases in organic carbon reported for Lake Gardsjon were higher than other increases reported in the literature. Broberg (1988) attributed this difference to the extensive sphagnum and benthic mats, which are somewhat unique to Lake Gardsjon. The literature suggests that liming does not significantly influence organic carbon levels in clearwater systems that do not contain extensive mats of sphagnum.

Toxic Metals

Metals that may be toxic to aquatic organisms—particularly aluminum, iron, lead, manganese, and zinc—are sometimes lower in limed waters due to at least four processes: precipitation of metals, oxidation, surface adsorption, and ion exchange. Typically, precipitation is the dominant process, because the solubility of metal ions is reduced in neutral or alkaline waters.

Liming is generally intended to reduce only the concentration of aluminum (among metals), and most studies report a significant reduction in aluminum concentrations after liming (Yan, 1983; White et al., 1984; Baalsrud, 1985; Wright, 1985; Marcus, 1988; Driscoll et al., 1989a; Gloss et al., 1989a). The major objective is a reduction in the toxic inorganic monomeric form of aluminum. Aluminum concentrations decrease in the water column because the solubility of $Al(OH)_3$ is dramatically reduced as pH increases from below about 4.5 to above 5. Baker et al. (1990b) gives a more detailed discussion of aluminum chemistry and Baker et al. (1990a) describes the effects of aluminum on aquatic biota.

Typical total aluminum concentrations in waters that are candidates for liming range from 200 to 1000 μg/L (Driscoll et al., 1980; Fraser and Britt, 1982). Liming to acidic, aluminum-rich waters causes hydrolysis and precipitation of the inorganic monomeric forms of aluminum. Driscoll et al. (1989a) observed significant alterations in the concentrations and speciation of aluminum after limestone treatment. Concentrations of inorganic monomeric aluminum decreased from about 250 μg/L to below 50 μg/L within 4 weeks in Cranberry Pond and 12 weeks in Woods Lake. No changes in organic monomeric aluminum were evident after liming. Wright (1985) observed similar responses of aluminum species for Lake Hovvatn in Norway.

Johannessen and Skogheim (1985) reported that liming caused the precipitation of inorganic monomeric aluminum, primarily as $Al(OH)_3$ or as $Al(OH)_4^-$ (aluminate), if the pH rises above 6.5. Further, Skogheim et al. (1986, 1987) demonstrated that at pH values above 8 aluminum bound to organic material is liberated and may form $Al(OH)_4^-$. Skogheim et al. (1986, 1987) also reported that the reduction in aluminum concentrations was greatest at the sites with the lowest concentration of organic matter in the water column.

There is some disagreement on the rate of reaction of aluminum precipitation. Dickson (1983) found that it normally takes 1–2 months for aluminum concentrations to decrease significantly when Swedish lakes are limed during extremely low water temperatures (i.e., 0–1°C). Wright (1985) observed similarly slow reaction rates for aluminum in Lake Hovvatn when the limestone was applied to the ice-covered surface, and neutralization occurred during ice melt when water temperatures were low. Driscoll et al. (1989a) found significant aluminum precipitation 4–12 weeks after liming.

Hasselrot et al. (1987), however, reported a relatively rapid decrease in aluminum concentrations after liming, and aluminum was precipitated rapidly in laboratory and enclosure experiments conducted by Johannessen and Skogheim (1985). The variations in reaction rates are probably due to variations in environmental factors that control precipitation, such as water quality variables, thermal conditions, and water turbulence.

Other metals that may be present in surface waters in concentrations toxic to aquatic organisms are sometimes significantly lower after liming. Table 3.3 summarizes the typical response of various metals to such treatments. The six studies showed only aluminum to be consistently lower in treated surface waters. Aside from aluminum, manganese and zinc were lower in four of the six studies. Driscoll et al. (1989a) reported that manganese was probably removed from the water column by direct precipitation. On the other hand, there was evidence that zinc was removed by adsorption or coprecipitation reactions.

None of the investigators reported significant changes in lead concentrations after liming. Iron concentrations actually increased in two of the studies after liming. White et al. (1984) attributed part of the increase to the iron content of the added limestone.

Many of these metals are present in surface waters in very low concentrations. This may explain why no decrease in some metal concentrations occurs after liming, even though solubility relations may suggest such changes.

Lake Sediment Responses

Liming often results in instantaneous absorption of calcium by lake sediments. This causes other changes in sediment chemistry and ion exchange reactions between the sediment and the overlying water. Many of these changes have been observed after liming. Unfortunately, the responses of lake sediments to liming have been documented in only a few studies: Lindmark (1982, 1985), Molot (1986), Gahnstrom (1988), and Driscoll et al. (1989a).

Some liming treatments have not resulted in significant changes in sediment chemistry. Driscoll et al. (1989a) observed little penetration of limestone through the water column to the sediments after treatment of Woods Lake and Cranberry Pond. Only minor sediment neutralization occurred after treatment. In other studies, particularly those in which application or large limestone particles was inefficient, liming was often followed by significant changes in sediment chemistry.

The changes that take place in the sediment and their effects on the chemistry of the overlying water are described in the next sections. Most studies involved the addition of limestone to the water column. One study is described on the effects of injection of sodium carbonate directly into lake sediments (Lindmark, 1982, 1985).

Base Neutralizing Capacity

The base neutralizing capacity, or alkalinity consumption, of

Table 3.3 Response of Selected Metals to Liming

Study	Response of Metal; Increase (+), Decrease (-), or Similar (o) Concentrations[a]					Reference
	Aluminum	Iron	Lead	Manganese	Zinc	
Living Lakes, N.E. United States	(-)	(o)	(o)	(o)	(-)	Marcus, 1988
Woods Lake and Cranberry Pond New York	(-)	(NR)	(NR)	(-)	(-)	Driscoll et al., 1989b
Middle and Hannah Lakes, Ontario	(-)	(+)	(NR)	(-)	(-)	Yan, 1983
Sandy Lake, Nova Scotia	(-)	(+)	(o)	(-)	(-)	White et al., 1984
Lake Hovvatn, Norway	(-)	(o)	(o)	(-)	(o)	Wright, 1985
Three lakes in Norway	(-)	(NR)	(o)	(NR)	(o)	Baalsrud, 1985

[a] NR = not reported.

sediments is a measure of the sediment-bound acidity that may be available to react with added base. It may influence the amount of alkaline material required to neutralize a lake, and sometimes decreases dramatically in lake sediments after treatment. The sediment chemical characteristics normally return slowly to preliming conditions, resulting in the potential contribution of alkalinity to the overlying water.

Molot (1986), who calculated the sediment base neutralizing capacity for several lakes on the basis of data in the literature (Table 3.4), found that the values for lakes with pH values of 4.2–7.4 varied between 0 and 3.5 mg $CaCO_3$/g wet weight. Unfortunately, no standard technique exists for measuring base neutralizing capacity; also the total amount in sediments is a function of both depth and chemistry of the sediment and interaction between the water column and interstitial water. The capacity often varies with depth; values are typically higher nearer to the sediment surface.

Sediment Chemical Changes

The dominant processes that occur in sediments after reaction with base materials are ion exchange reactions. For calcium-based materials, the dominant process is calcium ion exchange with H^+ bound to humic matter in the sediment (General Research Corporation, 1985). This results in increased pH and calcium content in the sediment. Sediment changes due to absorption of other dissolved cations such as magnesium are reportedly much less important (General Research Corporation, 1985).

Lindmark (1985), who measured the alkalinity released from the sediments of Lake Lilla Galtsjon after injection of the sediment with sodium carbonate, reported that over a 1-year period after treatment the net alkalinity release was 30 μeq/L. It was calculated that about 50% of the alkalinity in the sediments was consumed and exported. Although it was not reported, the reactions most likely involved sodium ion exchange with H^+ in the sediment.

Gahnstrom (1988) reported increases in pH of surface sediments and overlying water of Lake Gardsjon in the range of 1.5–2.5 units and 3–4 units, respectively. Wright (1985), on the other hand, observed little interaction of neutralizing material

Table 3.4 Estimates of Sediment Base Neutralizing Capacity Derived from Literature Data (Source: Molot, 1986)

Study	Analytical Technique	Collection Technique	Lake	Lake pH	mg $CaCO_3$/g Wet Weight		
DePinto and Edzwald, 1982	Incubated 1–10 g for 24 hr with 250 ml of water, initial alkalinity 2 meq/L.	Unknown	Big Moose	4.5	1.4–3.1		
			North	4.5	1.7–2.5		
			Nick's Pond	5.2	0.7–0.9		
			Ozonia	6.4	0.2–0.5		
Ripl and Lindmark, 1979	Dialyzed 70 ml of sediment for 15 days with 1 L of water (initial alkalinity 1.2–5.5 meq/L) and algae.	Ekman	Hinnasjon	6.1	1.8[a,f]	0.9[b,f]	[c,f]
			Fiolen	5.5	1.5	0.8	0.5
			Skarhultsjon	5.8	2.4	-	-
			Trummen	6.3	1.7	0.8	0.7
			Bergundasjon	7.1	3.5	1.4	0.8
			Lillesjon	6.4	2.7	-	-
Adamski and Michalski, 1975	Titration to pH 8.0 with $Ca(OH)_2$. Volumes and intermediate values not given.	0–20-cm core	Lohi	4.5	2.1 (0–8 cm)		
					1.7 (8–20 cm)		
			Middle	4.5	2.9 (0–8 cm)		
					2.2 (8–20 cm)		
Alenas et al., 1982	Titrated 20 mL of sediment to pH >8.0 with various amounts.	0–2-cm core	L. Harsjon	4.4–5.1	0.5[d,f]	1.8[e,f]	
			Blomman	4.5–5.3	0.3	0.9	
		1–10-cm depth	Langetjarn	4.2–5.3	0.3	1.0	
			Kullsjon	5.9–7.4	0.2	0.8	

Molot, 1982	Titrated 5 to 10 g dry wt. in 50 mL water with various amounts of $Ca(OH)^2$. Equilibrated for 3 days. Endpoint pH 6.4.	Ekman	Bowland	5.0	0.6–1.0
Dickson, 1980	Measured calcium (unknown technique) in 2-cm sections of core.	0–20-cm core	Horsikan	4.3	0.5 (0–2 cm)[f] 0.5 (2–4 cm) 0.4 (4–6 cm)
			0.3 (6–8 cm) 0.0 (8–10 cm)		
Hongve, 1977	Measured exchangeable Ca+Mg in 0 – 10 and 40- to 50-cm core sections with 1 *N* NH_4 acetate.	0–50-cm core	Svenskest-utjern	5.1	0.4 (0–10 cm) 0.0 (40–50 cm)
Oliver and Kelso, 1983	Measured exchangeable Ca+Mg in 1-cm sections of core with 1*N* NH_4 acetate.	0–50-cm core	Wavy	4.7	0.3 (0–1 cm)[f,g] 0.2 (1–2 cm) 0.1 (2–3 cm) 0.1 (3–4 cm) 0.0 (4–10 cm)

[a] Initial alkalinity 5.5 meq/L, endpoints pH 7.8–8.8.
[b] Initial alkalinity 2.2 meq/L, unknown endpoints.
[c] Initial alkalinity 1.2 meq/L, unknown endpoints.
[d] Endpoint pH 6.5.
[e] Endpoint pH 8.0.
[f] Water content of 90% assumed.
[g] Assumed average exchangeable Ca+Mg in sediments >4 cm was 145 μeq/g.

and sediments after limestone treatment of Lake Hovvatn and Pollen Pond.

Other reported changes in sediment chemistry after liming include increases in selected metals and changes in other constituents such as organic matter, nutrients, and sulfate.

Limestone treatment of White Deer Lake in Pennsylvania resulted in increased pH, heavy metals, and nitrate in the sediment during the first year after liming (Majumdar et al., 1987, 1989). There was also a decrease in organic matter, phosphorus, total nitrogen, ammonia nitrogen, sulfate, and aluminum. The pH of the lake sediment over the first year after liming increased from pH 4.9 to 5.4 to a high of 6.0. The sediment pH levels declined during the second year to levels of 5.7–5.8. The increases observed in cadmium, copper, lead, and zinc were reportedly due to the reduced solubility of these metals for the increased pH conditions in the sediment. The changes in nutrients and organic matter were attributed to the effects of increased decomposition rates in the less acidic sediments.

Sediment Oxygen Uptake

Gahnstrom (1988) studied the changes in sediment oxygen uptake after Lake Gardsjon was treated with limestone. The uptake decreased initially, then became significantly higher than pretreatment values, but declined to preliming levels after 15 months. It was suggested that this decline was due to oxygen uptake from the increased supply of organic matter in the sediments from increased decomposition (primarily sphagnum) and increased phytoplankton biomass, which resulted from liming. Other experiments with sediments, however, showed few significant changes in sediment oxygen consumption after liming, even though sediment chemical changes occurred.

EFFECTS ON BIOLOGICAL PROCESSES AND BIOTA

Biota in acidic surface waters generally respond positively to liming treatments. The results of biological studies of treated lakes and streams vary, however, due to a variety of factors, including

- preneutralization chemical and biological conditions,
- surface water hydrology and lake morphometry,
- timing of treatment and climatic conditions, and
- treatment strategies.

Further, positive responses in aquatic biota may be attributed not only to reduced physiological stress from improved water quality after liming, but improved food supplies or altered interspecific interactions between organisms (Weatherley, 1988).

The effects of liming on biological processes (nutrient cycling, decomposition, and primary productivity) and on biota (plankton, macrophytes, benthic macroinvertebrates, and fish) are discussed separately.

Biological Processes

In general, liming increases nutrient cycling, decomposition, and primary productivity. These observations have been particularly evident in lakes that were strongly acidic before liming or where base materials increased the pH of surface waters to relatively high levels (above pH 7.0).

Nutrient Cycling

Nutrient cycling typically increases in surface waters after liming. Increasing pH in sediments has been shown to cause the release of phosphate from the sediment to the water (Wright, 1985). The reported degradation of benthic sphagnum communities after the liming of certain surface waters results in the release of a substantial lake pool of organic carbon and organic nitrogen (Broberg, 1988). Sphagnum is often abundant in lakes that are strongly acidic; therefore, nutrient cycling would be expected to increase more markedly in those lakes after liming.

Weatherley (1988) reported that increased metal concentrations in sediments may limit the improved nutrient cycling reported for many liming studies. Metal complexes may bind phosphorus in sediment. Also, the toxic effects of metals on sediment bacteria may limit nutrient cycling. Little information exists on the relation between nutrient cycling and metal precipitation after liming (Andersson and Borg, 1988; Raddum et al., 1984).

Decomposition

Slow rates of decomposition of organic matter are typical for acidic lakes. The decomposition of plant detritus is crucial to the flow of energy and nutrients in most aquatic systems (Weatherley, 1988). Although the relation between decomposition and liming is sometimes inconsistent, most systems respond to increased pH and other chemical changes by increases in the numbers of sediment bacteria and resultant increases in the rates of decomposition (General Research Corporation, 1985; Majumdar et al., 1987, 1989; Gahnstrom, 1988).

Increases in decomposition are measured by changes in sediment oxygen uptake (Gahnstrom, 1988), breakdown of glucose and glutamic acid (Gahnstrom et al., 1980), decreases in concentrations of organic matter dissolved in sediments (Gahnstrom, 1988), and increases in nitrate (Majumdar et al., 1987). Short-term increases in decomposition may be due to the use of detritus built up during preliming acidic conditions (Weatherley, 1988). This may result in increases in nutrient cycling. These increases may or may not continue over the long term.

Little is known about the effects of metal precipitation on sediments and about the effects of increased metal content of sediments on decomposition. Scheider et al. (1975) observed decreases in bacteria in the sediments after lakes in the Sudbury area were limed. This may have been attributable to the abnormally high concentrations of metals in these lakes.

Primary Productivity

Liming often promotes primary productivity in previously unproductive waters by releasing phosphorus (usually the limiting nutrient to productivity in oligotrophic lakes) from sediments, as described previously. Another source of phosphorus that may result in increased primary productivity is the phosphorus content in the limestone. Relative to seasonal changes in nitrate before liming, Hornstrom and Ekstrom (1986) observed markedly reduced levels of nitrate during the first summer after liming, compared to fall and winter values, reportedly due to increased productivity. The authors attributed the increase in primary productivity to a combination of increased phosphorus supply from the sediments and the added phosphorus in the limestone.

The total annual primary production of different producers are shown in Table 3.5 for Lake Gardsjon, Sweden before and after liming. The total primary production in the lake was 1.5–3 times higher after liming due to the increased production of phytoplankton. Lower primary production was observed for epiphyton, benthic algae, and macrophytes, but these reductions were small compared to the change due to phytoplankton.

Plankton

Acidic surface waters generally have reduced the diversity of plankton species compared to that in circumneutral waters. Liming acidic surface waters results in changes in plankton species composition. Species diversity usually increases after liming. Liming often results in increased plankton biomass, but not in all limed surface waters.

Results of a variety of studies of the changes that have been observed in phytoplankton and zooplankton due to liming are next described to provide a broad range of effects for different surface water situations.

Phytoplankton

Only a few species of phytoplankton are typically present in acidic lakes. In southern Sweden, Eriksson et al. (1983) found only 5–20 species in strongly acidic lakes (pH <5.0), compared with 30–80 species in lakes with pH >6.0. Figure 3.1 shows the number of species of phytoplankton observed in May and August in relation to pH in several Swedish lakes.

Most phytoplankton species are eliminated in acidic lakes by high metal concentrations (e.g., aluminum), nutrient deficiency, or stress due to low pH (Eriksson et al., 1983; General Research Corporation, 1985; Weatherley, 1988). Baalsrud (1985) considered

Table 3.5 Total Annual Primary Production of Different Producers in Lake Gardsjon, Sweden, Before and After the Lake was Limed in 1982 (Source: Grahn and Sangfors, 1988)

	Total Annual Primary Production Per Unit Lake Area (g C/m²)	
Producer	Before Liming (1981)	After Liming (1985)
Phytoplankton	8.3	20–40
Epiphyton	1.2	<1.2
Benthic Algae	0.6	<0.6
Macrophytes	3.6	1.3
Total	13.7	20–43

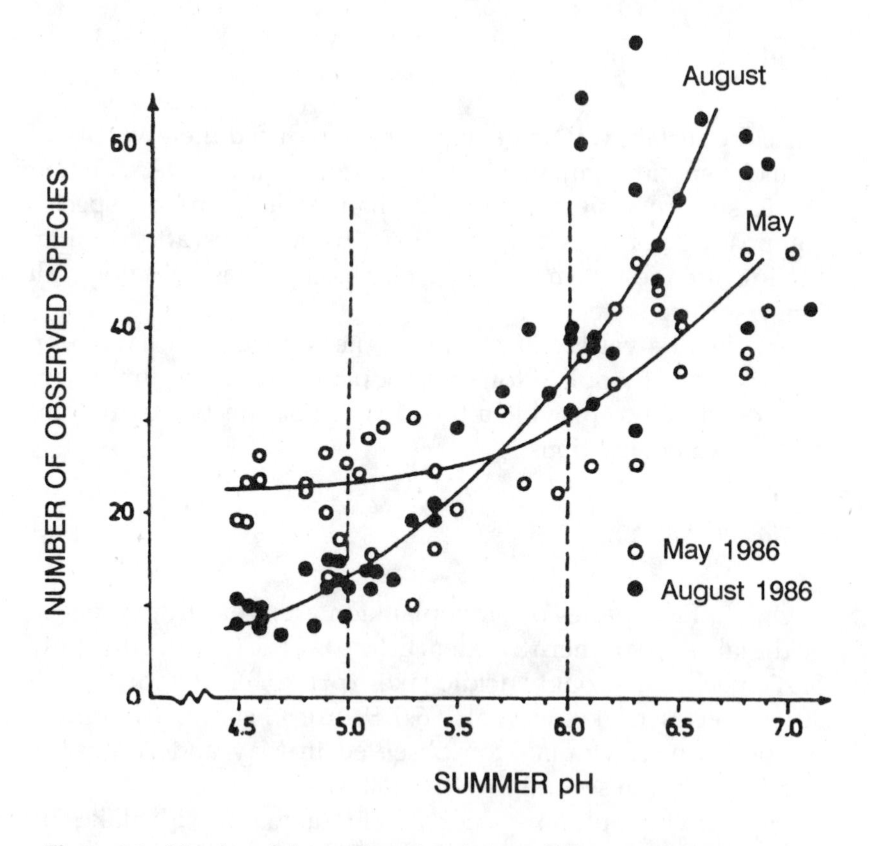

Figure 3.1 Number of observed species of phytoplankton in May and August 1976 in relation to pH in Swedish west coast lakes. (Modified from Eriksson et al., 1983.)

the low concentrations of inorganic carbon in acidic lakes to be an additional limiting factor. These factors also reportedly limit the abundance of phytoplankton biomass in more acidic lakes.

Most investigations of phytoplankton species composition after liming have revealed a transformation from a few acid-tolerant species to an increased number of species similar in composition to adjacent unlimed circumneutral oligotrophic lakes (Dillon et al., 1979; Hultberg and Andersson, 1982; Eriksson et al., 1983; Hornstrom and Ekstrom, 1986). Species composition generally shifts from a dominance by Dinophyceae (e.g., *Gymnodinium umberrimum* and *Peridinium inconspicuum*) to Chrysophyceae (e.g., *Chrysidiastrum catenatum*), Chlorophyceae, and Diatomaceae.

Many investigators have reported an initial reduction in phytoplankton biomass after liming, followed by gradual recovery to preliming levels and sometimes to higher levels (Adamski and Michalski, 1975; Bengtsson et al., 1980; Abrahamsen and Matzow, 1984; Siegfried et al., 1987). The initial reduction in biomass was due to pH shock resulting from an increase in pH that was either too rapid or too high immediately after liming. When no pH shock occurred, no phytoplankton dieback was reported.

The recovery of phytoplankton often depends on the nutrient status of the limed water body. Increased phosphorus in the water column (e.g., phosphorus released from the neutralizing material or the sediments) may enhance phytoplankton biomass from preliming levels. Weatherley (1988), however, stated that such effects are probably shortlived. Reduced toxicity, particularly due to lower H^+ and aluminum, has also been linked to increased phytoplankton production in limed lakes (Hornstrom and Ekstrom, 1986). For example, significant increases in phytoplankton biomass and number of species after limestone treatment of Lake Tvarsjon in Sweden were attributed primarily to a pH change from 4.8 to 7.4 and a reduction of over 50% in total aluminum.

Weatherley (1988) stated that grazing by herbivorous zooplankton sometimes influences phytoplankton populations. Thus, factors that control the abundance of these grazers indirectly affect phytoplankton biomass.

The following examples of phytoplankton response to liming provide an indication of expected responses for different situations (e.g., different water bodies and liming strategies).

Scheider et al. (1975) reported a severe decline in phytoplankton biomass after the liming of highly acidic Middle, Hannah, and Lohi lakes in Sudbury, Ontario. They attributed the reduction to the use of calcium hydroxide as the neutralizing material, which resulted in a rapid increase in pH. Phytoplankton biomass was reestablished within several months, but never surpassed preliming conditions. There was no change in phosphorus after liming. This lack of change was considered by Yan and Dillon (1984) to be the major factor that prevented phytoplankton biomass from increasing after liming. The composition of dominant species several months after liming had changed from Dinophyceae to non-acid tolerant Chrysophyceae, Cyanophyceae, and diatoms.

Bukaveckas (1989) observed significant reductions in phytoplankton density after liming. There was an immediate reduction in phytoplankton abundance immediately after three acidic Adirondack lakes were limed; however, no long-term deleterious effects were observed. At all sites, epilimnetic chlorophyll decreased by 50% within 3–4 days after treatment and after 10–11 days declined to less than 25% of pretreatment levels. Phytoplankton growth was suppressed for 24–25 days after liming. Rates of radiocarbon uptake during this period were less than 0.5 mg $C/m^3/hr$ and were only slightly greater than dark bottle values and zero time blanks. The period during which phytoplankton growth was suppressed corresponded to the period when the partial pressure of CO_2 was highly undersaturated with respect to atmospheric equilibrium. It is likely that the pretreatment phytoplankton assemblage was dependent on CO_2 as an inorganic carbon source for photosynthesis; CO_2 is the predominant form of dissolved inorganic carbon at pH <5. Phytoplankton growth may have been carbon limited after liming due to the depletion of CO_2, despite elevated concentrations of carbonate forms of dissolved inorganic carbon. These results suggest that physiological capacities for using CO_2 vs. bicarbonate forms of inorganic carbon play an important role in determining the species composition of phytoplankton communities experiencing changes in acid-base chemistry.

Chlorophyll levels were measured by Bukaveckas (1988a) as an indication of changes in phytoplankton production. Summer chlorophyll levels increased by twofold to threefold after liming Cranberry Pond and Woods Lake. The pH of Woods Lake was maintained between 6 and 7 over a 4-year period, during which time chlorophyll concentrations were generally greater than before liming and compared to a nearby untreated reference lake. In contrast, Cranberry Pond reacidified within 1 year after liming; reacidification was accompanied by a decrease in chlorophyll concentrations to pretreatment levels.

Hornstrom and Ekstrom (1986) observed low diversity in phytoplankton communities and the presence of only strongly oligotrophic species for 11 highly acidic lakes in southwest Sweden. After treatment of the water with limestone, there was a marked increase in number of species, and the dominant composition shifted to more eutrophic forms. In Lake Ommern, for instance, liming increased pH from about 5.3 to 7.0 and resulted in significant increases in alkalinity and phosphorus. The number of phytoplankton species doubled a year after liming (from about 30 to 60). Phytoplankton biomass also increased dramatically after liming, increasing from 0.1 to 0.6 mm^3/L two summers after liming. Diatoms such as *Asterionella formosa* dominated the biomass, whereas the previously frequent Dinophyceae were less prevalent.

Yan and Dillon (1984) reported that the addition of calcium hydroxide and limestone to slightly acidic Nelson Lake in Sudbury, Ontario increased the pH from 5.7 to 6.4. Because the lake was not highly acidic and contained some alkalinity, its phytoplankton community was not typical of nearby highly acidic lakes in which Dinophyceae dominated (Scheider et al., 1975). Before liming, the community was dominated by Bacillariophyceae, Chrysophyceae, and Chlorophyta. After liming, there was no significant change in species composition or biomass. Liming also had no effect on the temperature or oxygen structure of the lake or its nutrient content. It is likely that the Nelson Lake phytoplankton community was not affected by the liming, because preliming conditions were not highly acidic, the liming raised the pH only moderately, and the nutrient status of the lake did not change after liming.

Before liming Lake Gardsjon in Sweden, the phytoplankton community structure was dominated by large dinoflaggelate species such as *Peridinium inconspicuum* and *Gymnodinium* spp. (Larsson, 1988). After liming there was an initial decline in the number of species and biomass; the phytoplankton species then steadily increased in diversity and biomass. The trend continued until 1985 when a massive bloom of *Cosmocladium perissum* occurred. One possible explanation given for the bloom was that this species is not grazed by zooplankton and also has low nutrient requirements compared to those of many other phytoplankton species.

Zooplankton

There is an intricate relation between phytoplankton, zooplankton, and fish response in lakes that affects zooplankton response in limed lakes. Zooplankton is often the intermediate trophic level between phytoplankton and fish. Zooplankton is a particularly important food supply for the juveniles of many fish species (Weatherley, 1988). Zooplankters are also prey for benthic organisms. Zooplankton response to liming, therefore, often depends on whether fish are introduced into the lake.

Zooplankton species most tolerant to acidic conditions are the planktonic Rotifera and Copepoda (Bengtsson et al., 1980; Eriksson et al., 1983; Hasselrot et al., 1984; Hasselrot and Hultberg, 1984). In the absence of fish predation, liming generally results in increases in the abundance of these taxa within a year after liming (Eriksson et al., 1983). In addition, Henriksen et al. (1984) found that certain taxa often dominant at low pH are replaced by taxa that are less tolerant of acidic conditions. They observed a decreased population density of *Bosmina* spp. (Cladocera) and an increase in *Diaphanosoma brachyurum* (Cladocera).

Some zooplankton species may outcompete others after liming, and the population of one species may decrease at the expense of the other. For example, Brettum and Hindar (1985) attributed a decrease in the number of rotifers in Lake Hovvatn, Norway 1 year after liming and attributed the decrease to effective competition and eventual dominance by *Diaphanosoma brachyurum* (Cladocera).

Changes in zooplankton abundance and species composition in the absence of fish predation has been directly attributed to changes in water chemistry and increases in phytoplankton productivity and diversity after liming (Eriksson et al., 1983; Hasselrot and Hultberg, 1984; Henriksen et al., 1984; Hasselrot et al., 1987; Shortelle and Colburn, 1987). Weatherley (1988) reported that high concentrations of aluminum and hydrogen ions are known to be toxic to certain species of zooplankton through the effect on ionic regulation. Increases in phytoplankton production after liming provide increased food for herbivorous species of zooplankton when the species composition of the phytoplankton is suitable.

In situations where fish are reintroduced after liming or where some resident fish species are already present, the successful recruitment of fish species may result in predation and decreased abundance of zooplankton by juvenile fishes. Also, fish fry may change the make-up of the zooplankton community by selective predation on crustacean zooplankton that competes with rotifers for food (Henrikson et al., 1984; Shortelle and Colburn, 1987; Siegfried et al., 1987; Schaffner, 1989). Eriksson et al. (1983) reported that once fish were introduced into several limed lakes in Sweden, zooplankton dominance changed from Copepoda to Cladocera (e.g., *Daphnia* sp.). After liming Lake Stensjon, Henrikson et al. (1984) reported a substantial decrease in the abundance of all crustacean zooplankton. They attributed the decreases to intense predation by fish fry. Nyberg (1984) observed the elimination of chaoborid populations due to the recovery of fish populations after the neutralization of a Swedish lake. This elimination allowed the establishment of Cladocera as the dominant zooplankton crustacean.

Hornstrom and Ekstrom (1986) observed the most marked changes in zooplankton communities in lakes that had been strongly acidic before liming. For example, the number of zooplankton species in Lake Stora Harsjon, which had a preliming pH of 5.0, increased from less than 10 to nearly 20 some years after liming (Figure 3.2).

The neutralization of acidic waters is not always favorable to all zooplankton species. Of four Swedish lakes studied, Hornstrom and Ekstrom (1986) found liming unfavorable to some zooplankton species in three of the lakes. Only in Lake Tvarsjon was liming favorable to all species (Table 3.6). The disappearance of some zooplankton species was believed to be due to their preference for oligotrophic waters. The increased trophic level of the limed waters may have been responsible for the disappearance of *Heterocope* and *Bythotrephes*. Predation by planktivorous fish may also have played a role, particularly for *Bythotrephes* in Lake Stora Harsjon. Fish reproduction increased after liming.

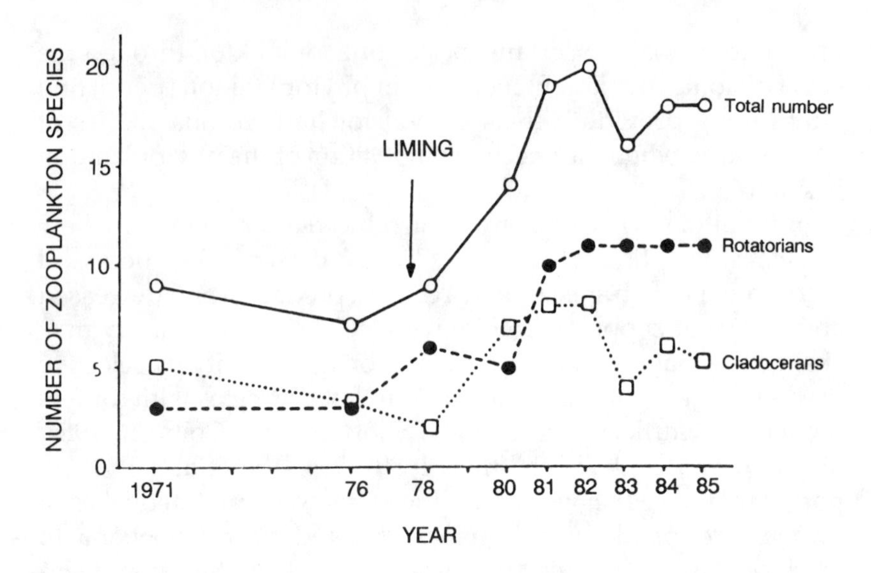

Figure 3.2 Number of zooplankton species in Lake Stora Harsjon, Sweden. (Source: Hornstron and Ekstrom, 1986.)

Macrophytes

Macroflora in aquatic ecosystems are an important energy and nutrient source and provide an important habitat for fish and other aquatic organisms (Weatherley, 1988). The macrophytes common to acidic lakes are the mosses, particularly sphagnum (Eriksson et al., 1983; Stokes, 1986; Grahn and Sangfors, 1988; Erikkson, 1988). Sphagnum, together with filamentous green algae (*Mougeotia* spp.) and benthic mats of blue-green algae, often form felt-like mats on the lake bottom. The mats sometimes cover and eradicate original species as lakes become more acidic.

These mats of sphagnum are greatly reduced in growth and distribution after liming (Hultberg and Andersson, 1982; Eriksson et al., 1983; Lindmark, 1985; Brettum and Hindar, 1985; Grahn and Sangfors, 1988; Erikkson, 1988). In some cases, the sphagnum mats were almost totally eradicated. For example, total macrophyte biomass and net production in Lake Gardsjon, Sweden were reduced by about 50% and 70%, respectively, after liming, mostly due to the disappearance of sphagnum (Grahn and Sangfors, 1988).

Clymo (1973) showed that a combination of high pH (>6.0) and high calcium concentration (>10 mg/L) is lethal to most species of sphagnum. Most liming treatments establish these conditions.

Table 3.6 Zooplankton Species Favored (+) and disfavored (-) by Liming in Four Southwest Swedish Lakes (Source: Hornstrom and Ekstrom, 1986)

	Lake			
Species	Tvarsjon	St. Harsjon	O. Nedsjon	Ommern
Ciliata		(+)	(+)	(+)
Asplanchna priodonta		(+)	(+)	
Conochilus unicornis	(+)	(+)		(+)
Gastropus stylifer		(+)		
Kellicottia longispina	(+)	(+)		
Keratella cochlearis	(+)	(+)		
Keratella quadrata		(-)	(-)	(-)
Keratella serrulata		(-)	(-)	(-)
Ploesoma truncatum		(+)	(+)	
Polyarthra vulgaris	(+)	(+)		(+)
Synchaeta sp.		(+)	(+)	(+)
Bythotrephes longimanus		(-)	(-)	
Ceriodaphnia quadrangula	(+)	(+)		
Daphnia cristata	(+)	(+)	(+)	(+)
Diaphanosoma brachyurum	(+)	(+)		
Holopedium gibberum		(-)		(-)
Cyclops spp.	(+)	(+)		
Eudiaptomus gracilis		(-)	(-)	
Heterocope appendiculata			(-)	(-)

Because of their strong ion exchange capacity and absorption of sediment nutrients, loss of the sphagnum community after liming may increase the availability of nutrients and ions important to biological productivity (Hultberg and Andersson, 1982). Loss of these groups may also enhance invertebrate habitat (e.g., isoetids such as *Lobelia* and *Isoetes*) by allowing colonization of newly exposed areas (Weatherley, 1988). Grahn and Sangfors (1988), however, did not observe recolonization of areas previously covered by dense sphagnum mats.

With the exception of species of sphagnum, few changes of quantitative importance in macrophyte distribution, biomass, or production have been observed after liming. Of four Swedish lakes studied, Eriksson et al. (1983) found consistent changes in macrophytes after liming only for sphagnum, with decreases in both distribution area and density in all lakes (Table 3.7). A few

Table 3.7 Response of Macrophyte Species to Liming in Four Southwest Swedish Lakes (Source: Eriksson et al., 1983)

	Lake			
Species	Trehorningen	Langsjon	St. Sirsjon	V. Skalsjon
Helophytes				
Carex rostrata	(0)	(0)	(+)	(0)
Carex lasiocarpa	(0)	(0)	(+)	(0)
Carex vesicaria		(0)		
Equisetum fluviatile	(0)	(0)	(0)	(0)
Eleocharis palustris		(0)		
Menyanthes trifoliata	(0)		(0)	(0)
Typha latifolia		(0)		
Lythrum salicaria		(0)		
Lysimachia thyrsiflora		(0)	(0)	(0)
Nymphaedids				
Glyceria fluitans		(0)		
Potamogeton natans	(*)	(*)	(0)	
Nymphaea alba	(+)	(0)	(0)	(0)
Nuphar lutea	(+)	(0)	(0)	(0)
Sparganium spp.	(+)	(+)	(0)	(0)
Eloeids				
Myriophyllum alterniforum		(+)	(0)	
Juncus bulbosus		(0)	(-)	(0)
Utricularia vulgaris	(+)		(0)	

Isoetids				
Lobelia dortmanna		(0)	(0)	(0)
Isoetes lacustris		(0)	(0)	(0)
Littorella uniflora		(0)	(0)	(0)
Ranunculus flammula		(0)		
Mosses				
Drepanocladus trichophyllus		(0)		(+)
Drepanocladus sp.	(0)		(0)	(0)
Fontinalis antipyretica		(0)		(0)
Fontinalis dalecarlica		(0)		
Sphagnum subsecundum (coll.)	(-)	(-)	(-)	(-)

(0) = Unchanged distribution area and density.
(+) = Increased distribution area or density after liming.
(-) = Decreased distribution area or density after liming.
(*) = New species after liming.

other species increased in distribution and density after liming in some lakes. One other species (*Juncus bulbosus*) decreased, but only in one of the four lakes.

The response to liming of the dominant macrophyte species (*Utricularia purpurea*) in Woods Lake and Cranberry Pond in the Adirondacks was evaluated by Bukaveckas (1988b). This macrophyte is often a dominant component of the macroflora of acid sensitive and acidic lakes in the northern United States and Canada. A progressive dieback in the species followed limestone treatment. The likely causes cited for the dieback were the initial reduction in CO_2 and the reduced transparency of the nonhumic lake after liming.

Dodge et al. (1988) reported no change in the sphagnum community of Bowland Lake, Ontario after liming. Extensive mats of sphagnum were not present in this lake before liming. Mats of filamentous green algae were present in the lake before liming; the mats disappeared after limestone treatment (Jackson et al., 1990).

Weatherley (1988) suggested that macrofloral growth is inhibited by shading when liming causes high epiphytic growth and phytoplankton blooms. On the other hand, macrophyte growth may be enhanced when humic lakes are limed and increase in transparency.

Lazarek (1986) evaluated shading and the relation between the macrophyte *Lobelia dortmanna* and changes in epiphyton and phytoplankton due to liming. Even though the biomass of phytoplankton increased as a result of liming, dramatic changes in macrophyte shading were not seen because the biomass of epiphyton was reduced.

Benthic Macroinvertebrates

The benthic organisms of lakes and streams are useful indicators of ecosystem status and are important prey for many fish. The benthic fauna of acidic lakes are often dominated by acid-tolerant species, such as the chironomids.

The community structure of benthic invertebrates may be considerably altered as a result of liming. The response of benthos to liming varies due to complex interactions between chemical and biological factors. There is no general relation between the response of benthic species to liming and preliming chemical and biological conditions in surface waters.

In some studies, the abundance of benthic invertebrates initially declined after liming (Scheider and Dillon, 1976; Hultberg and Andersson, 1982). This decline is possibly due to reduced sedimentation of organic material, an important nutrient for benthic organisms—particularly chironomid larvae (Eriksson et al., 1983).

Several taxa present at the time of liming usually increase in abundance (Hultberg and Andersson, 1982; Eriksson et al., 1983; Henrikson and Oscarson, 1984; Booth et al., 1987; Hasselrot et al., 1987; Bradt et al., 1989). Detritivores such as chironomids, oligochaetes, ephemeropterans, and the isopod *Asellus aquaticus* are often more abundant due to the increased microbial activity that usually accompanies liming. Sometimes, however, acid-tolerant chironomids are replaced over a period of 1 or 2 years by non-acid-tolerant species. Shortelle and Colburn (1987) considered fish predation to be a likely cause of such a change due to stocking and successful recruitment after liming.

Eriksson et al. (1983) reported little change after liming in species of the orders Odonata, Trichoptera, and Megaloptera. The decreases that did occur were believed to be caused by an increased intensity of predation by fish. For example, populations of *Sialis lutaria* increased after liming in lakes without fish, but not in lakes with fish.

Fjellheim and Raddum (1988) observed no significant differences in leaf processing by benthic invertebrates after the liming of Lake Hovvatn, Norway. The investigators attributed the low response to the densities of shredders, which remained low after liming.

Increases in the abundance of benthic fauna are usually more pronounced in littoral than in profundal zones of lakes (Eriksson et al., 1983; Hasselrot et al., 1987). Eriksson et al. (1983) found a dramatic increase in development of the benthic community in the littoral zone 2 years after nine lakes were limed in Sweden (Figure 3.3). An average 400% increase in the total number of benthic organisms was observed in these lakes. The total number of benthic organisms in the profundal zone increased an average of 250% during the first season after liming and then ranged between 127% and 175% of preliming values.

Bradt et al. (1989) reported that it took 2–3 years after liming

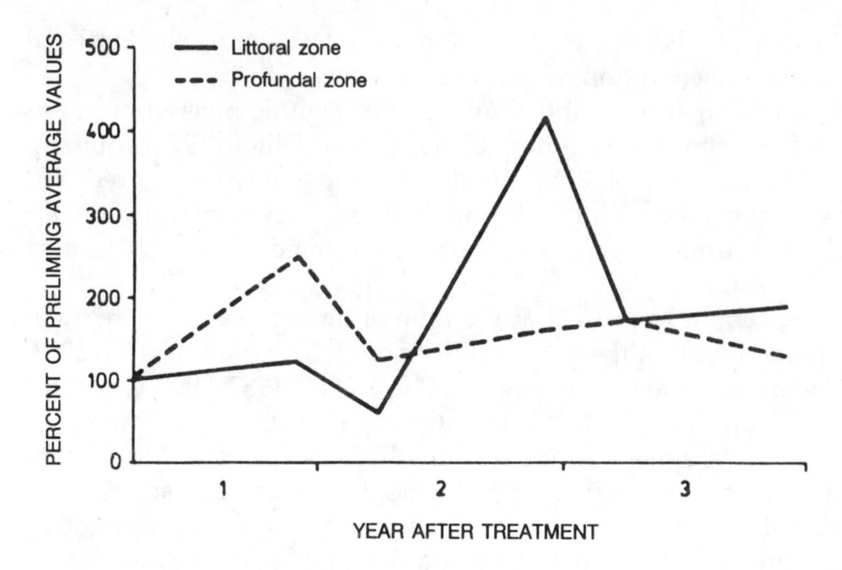

Figure 3.3 Development of the benthic community after liming, reported in percent of preliming average values for nine Swedish lakes. (Source: Eriksson et al., 1983. With permission.)

White Deer Lake, Pennsylvania for some of the acid sensitive invertebrates to become established. Increases in total numbers, wet weight, and taxa richness were observed in Ephemeroptera, Amphipoda, and Oligochaeta.

In some studies the number of benthic species in the profundal zone decreased after liming (Raddum et al., 1984; Brettum and Hindar, 1985). The decreases were attributed to changes in factors such as temperature, oxygen conditions, or precipitation of metals (e.g., aluminum) in the profundal areas. Build-up of metals in the sediment should be greater in the deeper portions of a lake (Weatherley, 1988). Reductions in benthic fauna due to aluminum precipitation in limed streams were also observed (Weatherley, 1988), but they occurred in streams of high ionic strength affected by acid mine drainage.

Henrikson (1988) observed negative effects to benthic fauna in littoral areas of Lake Gardsjon after liming because of the inflow of acidic, aluminum-rich water during storm episodes. One would not expect short-term storm events to affect benthos if the watershed was limed, instead of the lake water itself. Similar negative effects to benthic fauna in littoral areas were reported by Booth et al. (1987) for Bowland Lake, Ontario, but the cause was attributed solely to fish predation. Lake trout were introduced after liming, and the density of lake trout and yellow perch were highest in the littoral areas of the lake.

Weatherley (1988) suggested that interspecific interactions between benthic macroinvertebrates may be an important determinant of changes in species diversity and abundance. Predators may be affected by both water quality and the recovery of prey populations. An increased abundance of species present at the time of liming could intensify competition and inhibit reinvasion by other species. These interactions, however, have received little attention with respect to liming effects.

Fish

A reduction in numbers and the extinction of fish populations has been noted in lakes that have become acidic over time. The major reason for these occurrences is the mortality of the early life stages of fish, particularly eggs and larvae (Baker et al., 1990a).

The effects of liming on fish have received more attention than the effects on other aquatic biota because of the commercial, recreational, and resource management concerns related to fish survival, growth, and reproduction. The effects of liming have generally been favorable to fish populations. Improvements in the water quality of aquatic systems through liming permits the stocking of fish species previously lost from the system, introduction of new fish species, or the recovery of existing but stressed fish populations.

Beneficial Effects of Liming on Fish

Investigators in several countries have indicated beneficial effects with many different species of fish, both resident and introduced.

Many acidified lakes and running waters have already lost fish populations that were previously present, necessitating the reintroduction of fish species after liming. The most common species introduced into limed waters are lake trout, brook trout, Arctic char, rainbow trout, brown trout, Atlantic salmon, and smallmouth bass.

Various fish species have been shown to benefit from liming in lakes (Table 3.8) and running waters (Table 3.9). The effects of liming on the reproduction and growth of naturally occurring and introduced fish species is reviewed in the following sections.

Effects on Reproduction

Liming has been shown to restore fish reproduction in lakes and running waters with residual fish stocks where recruitment was sporadic or no longer occurred (Eriksson et al., 1983; Nyberg, 1984; Nyberg et al., 1986; Degerman and Nyberg, 1989). This restoration was usually the result of enhanced survival of early life stages, especially larvae, which are often considered the most sensitive to acidity-related factors (Weatherley, 1988). The possibility that liming may also restore female fecundity was suggested by Weatherley (1988), who noted that acidic conditions have been shown in the laboratory to disrupt female reproductive physiology.

Successful reproduction occurred in two Swedish lakes, even though the resident fish before treatment were few and very old (Eriksson et al., 1983). Soon after liming Lake Mortsjon, the youngest roach caught were 9 years old. Reproduction occurred the next spring and test fishing 3 years later yielded a large number of 1- and 2-year old roach. Just before liming, the perch population in Lake V. Skalsjon was almost eliminated, consisting of only about 10 mature individuals. One year after liming offspring were produced, and recovery has reported to be continuing (Nyberg, 1984).

Gunn and Keller (1980, 1981) reported that in acidic George Lake, southwest of Sudbury in the LaCloche Mountains, the *in situ* incubation of eggs of rainbow trout, lake trout, and brook trout within crushed limestone substrates greatly enhanced hatching success and sac fry survival during 7- and 30-day holding periods. A potential problem may be the precipitation of metals over the limestone surfaces or the acclimation of fry on the transition to ambient lake water (Weatherley, 1988).

Gunn et al. (1988) reintroduced smallmouth bass into Nelson Lake in Sudbury, Ontario after liming in 1975 and demonstrated the reestablishment of a reproducing population. Although the reestablished population was considered a direct result of improved water quality, the investigators suspected that water quality in the lake began to improve before liming because of decreasing emissions from the nearby Sudbury smelters.

Brown trout were stocked in Loch Fleet, Scotland after watershed liming and survival was successful (Brown et al., 1988). Trout fry were observed the year after liming in both inlet and outlet streams, indicating that successful reproduction had occurred.

Successful reproduction of brown trout and Atlantic salmon has been demonstrated after liming of river systems in Canada, Sweden, and Norway (Watt et al., 1984; White et al., 1984; Eriksson et al., 1983; Nyberg, 1984; Rosseland et al., 1986; Rosseland and Hindar, 1988; Appelberg et al., in press). Watt et al. (1984) and White et al. (1984) reported that liming Sandy Lake in Nova Scotia controlled the low pH in Atlantic salmon habitat downstream from the lake. In the Tjostelrodsan River, Sweden, the numbers of under yearling and yearling brown trout were extremely low during the years before liming in 1975 (Nyberg, 1984). In 1975, the upstream lake was treated with limestone and the water quality in the river improved. Electrofishing in 1976–1980 yielded significantly higher numbers of yearlings and older trout than were taken in the 4 years before liming (Figure 3.4). Survival experiments were performed with 50 Atlantic salmon smolts just before the River Audna in Norway was limed (Rosseland and Hindar, 1988). All died within 6 days. In 1985, continuous liming with automated dosing devices was begun in the river. Salmon were then stocked and spawning was confirmed by 1987.

Effects on Growth

The effects of liming on the growth of fishes depends on the status of existing fish populations before liming (Nyberg, 1984; Hasselrot and Hultberg, 1984; Nyberg et al., 1986). Fish usually grow well in waters with no recent fish reproduction because of the lack of competition for food. After treatment and recovery of recruitment, growth sometimes declines due to increased competition for food. However, Nyberg et al. (1986) and Degerman and Nyberg (1989) showed that growth generally increases after moderately acidic lakes are limed. They suggested that growth declines occur due to food reduction and physical stress only in

Table 3.8 Lake Fish Populations that Were Reported to Benefit from Liming

Species	Observed Effect after Liming	Lake	References
Arctic char	Reintroduction, spawning, and enhanced survival	Swedish lakes	Bengtsson et al., 1980; Eriksson et al., 1983; Hasselrot et al., 1984; Nyberg et al., 1986; Nyberg, 1987
Brook trout	Protection of early life stages by placing larvae in limestone-filled containers	George Lake, Ontario	Gunn and Keller, 1984
	Reintroduction, spawning, and enhanced survival	Swedish lakes	Hultberg and Andersson, 1982
	Reintroduction, spawning, and enhanced survival	Adirondack lakes, New York	Blake, 1981; Schofield et al., 1986; Simonin et al., 1988; Gloss et al., 1989a
	Introduction and enhanced survival	Spruce Knob Lake and Summit Lake, West Virginia	Fraser and Britt, 1982
Brown trout	Reintroduction, spawning, and enhanced survival	Swedish lakes	Hultberg and Andersson, 1982; Nyberg et al, 1986

	Reintroduction and enhanced survival	Norwegian lakes	Rosseland and Hindar, 1988
	Reintroduction, spawning, and enhanced survival	Loch Fleet, Scotland	Brown et al., 1988
	Reintroduction and enhanced survival	Two lakes in Wales	Welsh Water, 1986
Cisco	Enhanced survival	Lake Nedsjon, Sweden	Bernhoff, 1979; Degerman and Nyberg, 1989
European perch	Spawning and enhanced survival	Swedish lakes	Bengtsson et al., 1980; Hultberg and Andersson, 1982; Lindmark, 1985; Nyberg et al., 1986; Degerman and Nyberg, 1989
Lake trout	Protection of early life stages by placing larvae in limestone-filled containers	George Lake, Ontario	Gunn and Keller, 1984
	Enhanced survival of eggs and fry in limestone-treated spawning shoals	Bowland Lake, Ontario	Booth et al., 1987
	Enhanced survival and spawning	Nelson Lake, Ontario	Gunn et al., 1988
	Reintroduction and spawning	Bowland Lake, Ontario	Booth et al., 1987

Table 3.8 Lake Fish Populations that Were Reported to Benefit from Liming (continued)

Species	Observed Effect after Liming	Lake	References
Northern pike	Spawning and enhanced survival	Swedish lakes	Hultberg and Andersson, 1982
Rainbow trout	Protection of early life stages by placing larvae in limestone-filled containers	George Lake, Ontario	Gunn and Keller, 1984
	Introduction and enhanced survival	Spruce Knob Lake and Summit Lakes, West Virginia	Fraser and Britt, 1982
	Reintroduction, spawning, and enhanced survival	Adirondack lakes, New York	Kretser and Colquhoun, 1984
Roach	Spawning and enhanced survival	Swedish lakes	Bengtsson et al., 1980; Hultberg and Andersson, 1982; Eriksson et al., 1983; Lindmark, 1985, Nyberg et al., 1986; Degerman and Nyberg, 1989
Smallmouth bass	Reintroduction, spawning, and enhanced survival	Nelson Lake, Ontario	Gunn et al., 1988

Table 3.9 Fish Populations in Running Waters that Were Reported to Benefit from Liming

Species	Observed Effect after Liming	River or Stream	References
Atlantic salmon	Reintroduction, spawning, and enhanced survival	Swedish rivers	Appelberg et al., in press
	Enhanced survival	Hogvadsan River system, Sweden	Bengtsson et al., 1980; Nyberg et al., 1986
	Reintroduction, spawning, and enhanced survival	River Audna system, Norway	Rosseland et al., 1986; Rosseland and Hindar, 1988
	Enhanced survival	Sackville River system, Nova Scotia	Watt et al., 1984
	Enhanced survival	Nant Craflwyn, Wales	Weatherley et al., 1989
Brown trout	Reintroduction, spawning, and enhanced survival	Swedish streams	Eriksson et al., 1983; Nyberg et al., 1986
	Enhanced survival	River Audna system, Norway	Rosseland and Hindar, 1988
	Reintroduction, spawning, and enhanced survival	Tjostelrodsan River, Sweden	Nyberg, 1984
Yellow perch	Enhanced survival	Maryland streams	Janicki and Greening, 1988a,b

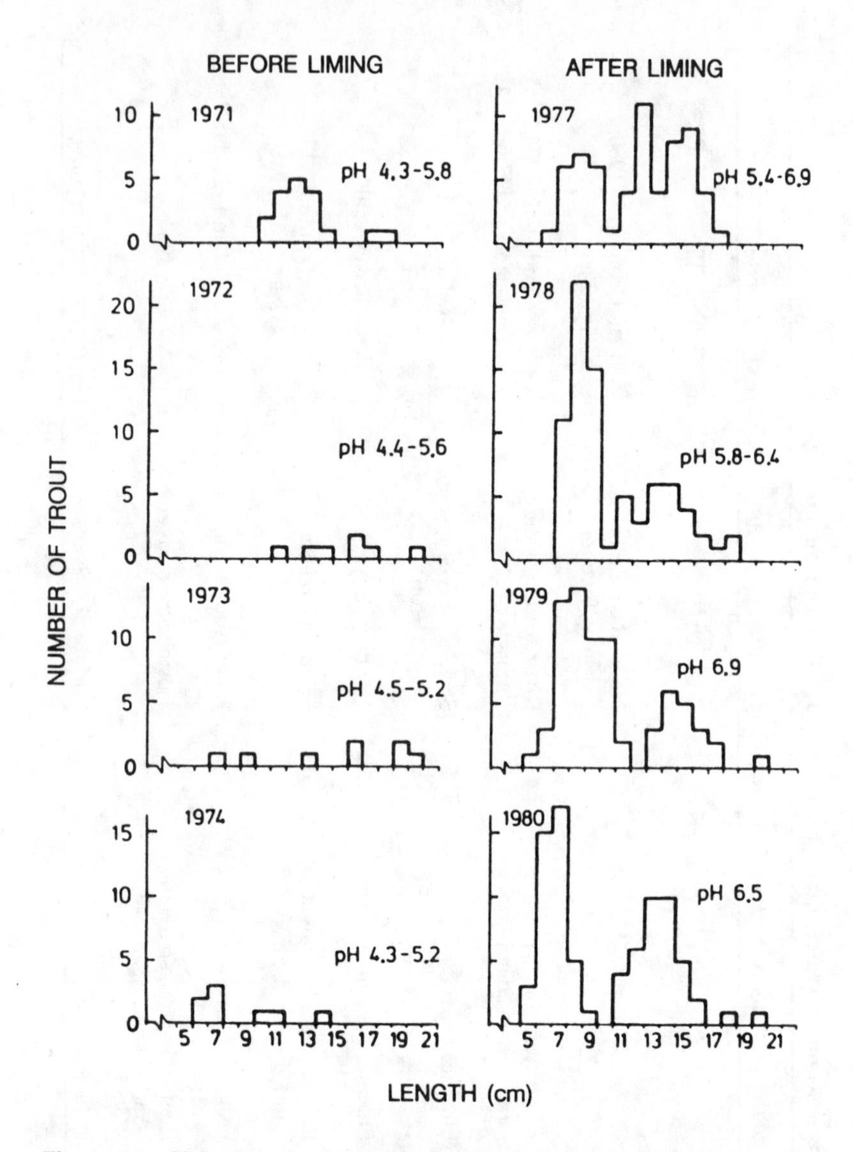

Figure 3.4 Changes in the stock of sea trout in the Tjostelrodsan River before and after liming the upstream lake. (Modified from Nyberg, 1984.)

severely acidified lakes. For example, growth of the few remaining European perch ceased after Lake Rodvattnet (pH 4.8) was limed (Degerman and Nyberg, 1989).

Improved growth of fish after liming is not the only important consideration. Liming is often conducted to increase sport fishing; thus, it is more important to increase total production or biomass.

A few fast-growing fish have less total biomass than numerous fish with a slower growth rate.

After the liming of Lake V. Skalsjon, Sweden, in 1975, recruitment to the perch population was again seen and growth was good for the first few years (Eriksson et al., 1983). Figure 3.5 shows the post-liming growth in fish 1 year (1976) and 5 years (1980) later and compares this growth to that in 1969. Growth was good 1 year after liming, but had declined sharply by 1980, though there was still some reproduction.

Depressed growth rates have been observed at higher standing crops and densities of fish in Adirondack lakes (Schofield et al., 1986; Gloss et al., 1989a,b; Schofield et al., 1989). Brook trout were reintroduced into several lakes in the Adirondack region of New York State after liming. The initial growth and condition of introduced fish were good in all lakes where liming produced suitable water quality. Higher stocking rates, however, were found to lead to increased foraging pressure on the limited food, resulting in depressed mean growth rates. Because long-term detailed fisheries studies have not been conducted on these lakes, it is not known whether a balance could be maintained between stocking and growth rate after 5 or more years of continued maintenance liming.

Fish Mortality Related to Liming

Liming has seldom caused mortalities in resident fish populations, but a few isolated incidents of mortality have occurred due to metal toxicity (Powell, 1977; Bengtsson et al., 1980; Flick et al., 1982; Dickson, 1983; Nyberg, 1984; Nyberg et al., 1986). These deaths have often been due to improper treatment and stocking of fish, or the treatment of lakes with high metal concentrations from local atmospheric sources (e.g., Sudbury lakes and Canadian smelters).

The pH-dependent toxicity of aluminum to fish was described by Driscoll et al. (1980), Schofield and Trojnar (1980), Baker (1981), Baker and Schofield (1982), and Baker et al. (1990a). Also, some aluminum species are more toxic to fish than others. For example, labile (inorganic) monomeric aluminum is much more toxic than nonlabile (organic) aluminum.

During the liming of a highly acidic lake, organisms may be

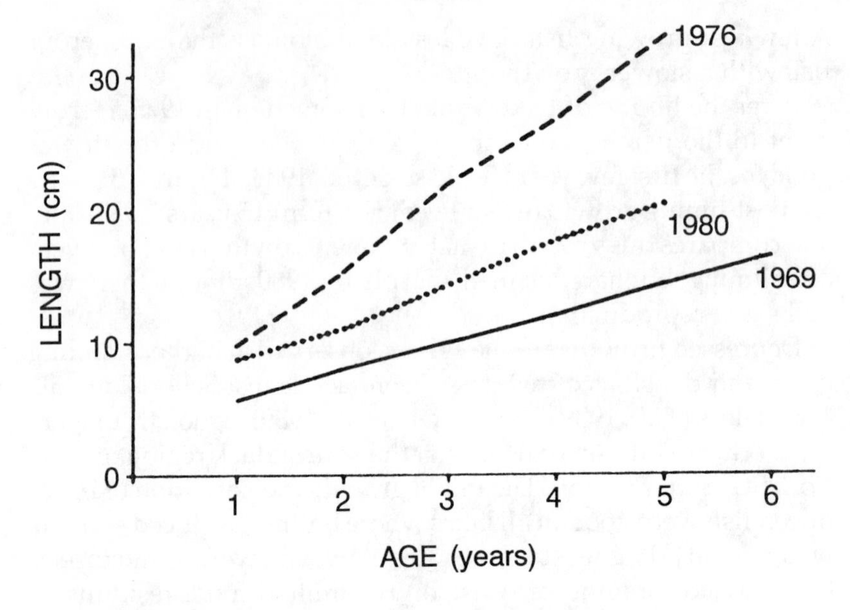

Figure 3.5 Growth of European Perch in Lake Skalsjon before liming (1969), 1 year after liming (1976), and 5 years after liming (1980). (Source: Eriksson et al., 1983. With permission.)

adversely affected during the transition period before metals have precipitated. It may be that during the elevation in pH, metal concentrations remain high and metal hydroxides precipitate onto gills, leading to death due to combined osmoregulatory and ventilatory stress (Weatherley, 1988). Toxicity might also occur due to increased concentrations of hydrated metal ions such as aluminate [$Al(OH)_4^-$]. Aluminate ions, which increase as pH increases, may cause toxicity.

Toxicity may be more likely when bases with monovalent cations (e.g., $NaCO_3$) are added (Rosseland and Skogheim, 1984). Additional protection to fish from toxic forms of aluminum results when the alkaline material is calcium based (Abrahamsen and Matzow, 1984). Seemingly, the calcium ions compete with metal ions for binding sites on the surface of the fish gill (Brown, 1983). This influences the uptake of these metals, and ultimately their toxicity to fish (Hunn et al., 1985).

In Middle and Lohi lakes, which are only about 5 km from the Sudbury smelters, highly elevated metal concentrations produced fish mortalities in the stocked species (smallmouth bass in Middle Lake and brook trout in Lohi Lake), even after whole-lake neutralization raised the pH (Powell, 1977; Keller et al., 1980). Powell (1977) suspected that the mortalities of stocked fish in the two lakes were related to copper toxicity; later experiments with rainbow trout in these two lakes confirmed this suspicion (Yan et al., 1979).

Aluminum toxicity reportedly killed salmon and rainbow trout stocked immediately after liming an acidic lake in Sweden (Fraser and Britt, 1982). Colloidal and particulate aluminum had not all precipitated when fish were stocked. The researchers suggested that mortality may not have occurred if the fish had been stocked after aluminum had completely precipitated.

Dickson (1983) reported mortality of rainbow trout after stocking Lake Ravekarrs Langevatten, Sweden. The mortality of rainbow trout and Atlantic salmon were also reported by Dickson (1983) in a fish hatchery receiving water from a limed upstream pond. In both systems, mortality was attributed to the colder waters, which slowed aluminum hydrolysis and precipitation.

Matzow (1985) reported symptoms of sublethal stress and some mortality of brown trout in enclosure experiments in Lake Bjordalsvatn, Norway. During a 6-day period after liming, the toxic labile (inorganic) monomeric aluminum fraction varied greatly.

In Woods Lake and Cranberry Pond, there was an unusually high concentration of aluminum immediately after liming, but there was no mortality in caged and stocked brook trout (Gloss et al., 1989b). The high concentration of nonlabile aluminum relative to total aluminum was considered an important factor in the survival of caged and stocked brook trout in the limed Adirondack lakes. In contrast, fish mortality followed liming of a lake in Sweden due to high concentrations of the more toxic labile (inorganic) monomeric form of aluminum (Dickson, 1983).

Gunn and Keller (1984) observed high concentrations of aluminum and iron hydroxides in stream water treated with instream limestone gravel, and suggested that the flushing of precipitates from the gravel bed during high flows might contribute to fish toxicity under some circumstances. In liming streams and rivers, it may be difficult to maintain acceptable water chemistry consistently due to power failure, equipment breakdown, and periods of high flow. During these times, the potential for adverse effects on fish are increased. Additional experience and potential improvements in the technology of liming running waters should maximize the maintenance of acceptable water quality and reduce the likelihood of impacts.

Although there have been no reports in the literature of fish mortalities due to acidic stream water entering a limed lake, the potential exists for mortalities to occur. Henrikson (1988) observed negative effects to benthic fauna in a limed lake because of the inflow of acidic, aluminum-rich water during storm episodes. Weatherley (1988) suggested that such risks may be small due to avoidance behavior in fish. Fish have been observed to seek refugia when exposed to adverse water quality (Haines, 1981). Avoidance, however, is possible only for mobile life stages of fish.

General

Variations in the effects of liming on fish may depend on the food supply at the time of liming. Nyberg (1984) observed that, in the early stages of acidification, many important fish food organisms, such as crustacean zooplankters, are extremely sensitive to acidification and may die before the fish themselves are directly affected. There may then initially be little food for fish that are stocked in a newly limed lake. On the other hand, in the later stages of acidification, the fish population may become sparse and the food supplies more abundant due to a decrease in predation. Stocked fish in this instance may initially have a large food supply.

The recovery of different fish species and the longer term success of stocking may depend on the various rates of recovery of other biotic components. The recolonization of fish food sources, such as invertebrates, after liming, may be exploited by resident or stocked fish populations. In Cranberry Pond, New York, the immediate effect of brook trout stocking was the near elimination of *Chaoborus americanus*, the dominant macroinvertebrate before liming (Gloss et al., 1989). Results from in situ enclosure experiments indicated that the decline in abundance could be attributed to increased brook trout predation rather than a physiological response to changes in lake chemistry (Robbins, 1990). The macroinvertebrate was the most prevalent in brook trout stomachs (Evans, 1989; Schaffner, 1989).

Prey-predator relations between different fish species may also affect their population levels. The increased supply of young perch, roach, and cisco after liming was observed to be a food source for predatory fish such as northern pike, European perch, and Arctic char (Nyberg, 1984).

Bukaveckas (in press) noted that liming often has a fertilizing effect on phytoplankton communities in Adirondack lakes. For lakes of moderate depth (10–15 m), increases in phytoplankton, coupled with stronger thermal stratification (due to reduced light penetration), may result in greater hypolimnetic oxygen depletion. Oxygen resources of the hypolimnion determine habitat suitability for cold-water fish species such as brook trout and lake trout. At Woods Lake, the oxygen saturation in summer before liming was 85%, but averaged only 54% during the five summers after liming. Higher rates of oxygen depletion resulted in partial water column anoxia in 3 of the 5 summers after liming, whereas oxygen concentrations were above 7 mg/L at all depths before treatment.

It has been postulated that surface waters susceptible to acidification (i.e., those with low acid neutralizing capacity and a high degree of physicochemical variation over rather short periods) normally have relatively marginal conditions for aquatic life (Weatherley, 1988). This has resulted in biota being adapted to these environments. One might expect that the stocking of "acid-tolerant" strains of fish would be more successful; however, Gloss et al. (1989b) found no significant difference between "acid-tolerant" and "normal" brook trout stocks introduced into Woods Lake and Cranberry Pond. They suggested that the rapid reacidification rate of these lakes precluded complete evaluation of the advantages of stocking fish considered to be acid tolerant.

Restoration of water quality conditions to those believed to exist before acidification has not always resulted in the restoration of the original biota. Weatherley (1988) suggested that the cause for this delay may be the need of many species for a longer period for colonization and the establishment of viable populations than has been evaluated to date. Degerman and Nyberg (1989) cited a few examples where it took about 10 years after liming for the fish population to return to a more normal distribution.

It may also be possible that liming cannot restore conditions exactly as they were before acidification. For example, whole-lake liming does not eliminate acidic episodes from influent streams or from littoral zones. Other factors mentioned previously also separate limed waters from their unacidified counterparts, including precipitated metals, undissolved base material, elevated calcium levels, and the possibility of reacidification between treatments.

Generally, however, the response of fish to liming is that most species affected by acidification respond positively and directly to the improvement in water quality that occurs after liming. Potential problems of gradual or sudden reacidification (from snowmelt or storm episodes) and later remobilization of toxic metals can be prevented. Watershed liming may be needed in some situations to prevent the episodic introduction of highly acidic water. When lakes that have been neutralized begin to reacidify and drop to a set pH or ANC level, the lake can be relimed before conditions detrimental to the aquatic life return.

4

Resource Management

"If people concentrated on the really important things in life, there'd be a shortage of fishing poles."

Doug Larson

MANAGEMENT OBJECTIVES

Mitigation of surface waters believed to be affected by acidic deposition has been conducted to attain one of two main objectives—restoration or prevention. Restoration liming is broadly defined as liming to improve the quality of water (principally to increase pH or ANC) in aquatic systems that are designated for a particular use, such as to support a fishery or serve as a drinking water supply. The term *mitigation* liming is often used interchangeably with *restoration* liming. Preventive liming is broadly defined as liming to maintain a reserve ANC in aquatic systems susceptible to the adverse effects of acidification. In some of the mitigation programs, the terms *maintenance* or *protection* are used instead of *prevention*.

In past mitigation programs, restoration liming was conducted primarily to improve water quality in lakes and streams that had once supported a healthy fishery. Current restoration programs (most commonly relying on liming the surface water or watershed) have a similar objective: to develop water quality suitable for the renovation of economically, recreationally, or ecologically important fishes (Nyberg and Thornelof, 1988; Brown and Goodyear, 1987; LLI, 1987). Candidate waters for which restoration treatments are appropriate might include those with pH <6 or ANC <0.

Preventive liming is conducted to maintain water quality conditions conducive to protecting an aquatic population or community deemed important from an ecological or recreational perspective, e.g., rare or endangered aquatic or amphibious species, fish populations that are valuable brood stock, natural heritage fish strains that are important for sport or commercial fishing, or a regionally unique species. Candidate sites may show declining or stressed fisheries resources or sufficiently low ANC to suggest that the continued maintenance of existing populations or communities is threatened. Signs of acid-related stress in fish include an increased incidence of recruitment failure, missing year classes, or loss of acid-sensitive species.

CANDIDATE SYSTEMS

Various procedures for selecting candidate surface waters for mitigation have been adopted for selected mitigation programs described earlier. Here are described site selection criteria, treatment strategies, target water quality objectives, and retreatment threshold values set by major mitigation programs. The number of acidic or acidified aquatic systems in the United States, Canada, Sweden, and Norway are estimated.

Criteria for Selecting Appropriate Surface Waters for Mitigation

Specific criteria for selecting surface waters suitable for mitigation by liming techniques were developed by six major mitigation programs: Living Lakes, Inc. (LLI), U.S. Fish and Wildlife Service, Massachusetts Division of Fisheries and Wildlife, New York Department of Environmental Conservation, and the Swedish and Norwegian government subsidy programs (Table 4.1). Criteria developed for each program are somewhat different, primarily because the programs were planned for different purposes. The LLI program was primarily a demonstration program, and the U.S. Fish and Wildlife Service program was for research. The remaining programs are operational and are the only ones where resource management is the sole objective.

Living Lakes, Inc.

The stated purpose of the LLI mitigation program is to demon-

Table 4.1 Major Site Selection Criteria for Lake and Stream Mitigation Programs

Program	Maximum pH	Maximum ANC (μeq/L)	Lake Retention Time (yr)	Lake Surface Area (ha)	Mean Monthly Stream Flow (m^3/sec)	Reference
Living Lakes, Inc.	6.0	10	≥0.5	≥2	≥50	LLI, 1987
U.S. Fish and Wildlife Service	6.0	10	0.5–6.0	≥5	≥25	Brown and Goodyear, 1987
Massachusetts	—	40	≥1	≥4	—	Massachusetts Division of Fisheries and Wildlife, 1984
New York	5.7	20	≥0.5	—	—	Simonin et al., 1988
Sweden	6.0	50	—	—	—	Nyberg and Thornelof, 1988
Norway	6.0	50	—	—	—	B.O. Rosseland, personal communication, 1990

strate the restoration or protection of important fisheries. An operations manual was developed that outlined inclusion requirements for lakes and streams (LLI, 1987). Many of the criteria were identical to those applied to site selection for the Fish and Wildlife Service program. The information presented in this section is a summary of the LLI program's requirements.

The pH and ANC of included sites must be ≤6.0 and ≤10 μeq/L (Table 4.1). Other criteria include lake retention time (≥0.5 year), lake surface area (≥2 ha), and mean monthly stream flow (≤50 m^3/sec). Separate criteria for restoration sites and maintenance liming sites are also defined (Table 4.2). Sites with insufficient background data to address the criteria presented in Table 4.2 are not considered for treatment unless they are subjects of site-specific research or the data required to evaluate their candidacy for inclusion can be easily obtained during a reconnaissance visit.

Inclusion criteria related to biological status (Table 4.2) differ for mitigative and maintenance liming projects. Mitigative liming projects must fulfill at least one of two conditions:

- a stressed fish population is present, or
- a historical record of a fishery in the system is documented, if no fish population or community is present.

Maintenance liming requires the presence of an ecologically or recreationally important fish population or community.

The pH and ANC criteria also differ for mitigative and maintenance liming projects: mitigative liming sites must have a pH ≤6.0 or an ANC ≤10 μeq/L, and maintenance liming sites must have a pH ≤6.5 or an ANC ≤50 μeq/L. The pH threshold for mitigative liming was based on the results of numerous field and laboratory investigations of fish response to acidic conditions. Acid-related stress (e.g., recruitment failure, absence of year classes, loss of acid-sensitive species) was observed in these studies at pH values of 5.8 or below for the following fish species: lake trout, rainbow trout, brown trout, and brook trout; smallmouth bass and striped bass; walleye; blueback herring; Johnny darter and Iowa darter; common shiner; and fathead minnow and bluntnose minnow. The selected pH threshold for maintenance liming is sufficiently high to allow potentially stressful water quality conditions to be avoided, that is, exposure to acidic conditions could be prevented before it occurred.

The maximum ANC criterion for mitigative liming (10 µeq/L) was selected as representative of conditions in which the water body has virtually no buffering capacity and pH is subject to extreme fluctuation. Baker (1984) stated that biological effects due to surface water acidification begin when aquatic ecosystems reach ANC values of 10–90 µeq/L.

The criteria for critical habitat, pollutant discharge, hydrology, and lake size are identical for mitigative and maintenance liming projects. Except for hydrology, the criteria apply equally to lakes and streams. The hydrology criterion is based on hydraulic retention time for lakes and the flow rate for streams.

The LLI program allows for any criterion to be waived in the selection of a research site as a candidate for liming. For example, as a research project, a lake with a hydraulic retention time of ≤0.5 year could be treated by using stream dosing technology.

The chemical neutralization of naturally acidic bogs is not considered a principal objective of the program, even if such systems have significant sport fishing potential; the liming of some organically colored waters, however, has not been precluded under the LLI program. In several of the ponds treated in 1986 and 1987, dissolved organic carbon concentrations exceeded 5 mg/L (Marcus, 1988).

U.S. Fish and Wildlife Service

The U.S. Fish and Wildlife Service initiated the Acid Precipitation Mitigation Program in 1985. This cooperative research program with the states produced a two-volume report that included criteria for selecting candidate sites for inclusion in the program (Saunders et al., 1985; Fraser et al., 1985; Brown and Goodyear, 1987). Brown and Goodyear (1987) defined several criteria to screen and select candidate lake and stream research sites, including programmatic, hydrological, and research criteria, as well as the consideration of confounding factors (summarized in Figure 4.1).

Specific criteria for including a site in the program follow:

- The geographic region currently must be experiencing acidic deposition (i.e., an annual average rainfall pH <5.6), and this must be the only major factor posing a risk to the system.
- Surface water pH should be less than 6.0 (or alkalinity ≤10

Table 4.2 Matrix Describing Site Selection Criteria Adopted Under the Living Lakes Liming Program (Source: LLI, 1987)

Type of Project	Site Selection Criteria[a]					
	Biological Status	pH/Alkalinity	Critical Habitat	Pollutant Discharges	Hydrology	Lake Size
Mitigative liming	The system must have an existing fish population (naturally reproducing, put-and-take, or put-grow-and-take), or it must have historical records of fish in the system. If fish are present, deleterious effects should be indicated (i.e., increased incidence of recruitment failure; missing year classes; loss of acid-sensitive species). Such stress should not be attributable to nutrient enrichment, point source pollution, interspecific competition, or fishing pressures.	Surface water pH values ≤6.0 or acid neutralizing capacity (ANC) values ≤10 μeq/L should exist in some significant spatial location in the system as determined sometime during the past 5 years. Such conditions should either (1) occur frequently, (2) be maintained for 2 or more weeks, or (3) occur during critical periods in sensitive life stages of important aquatic biota.	Suitable habitat (spawning sites, nursery areas, shelter and dissolved oxygen and temperature regimes, etc.) must exist for the species of interest. If stress has been documented for acid-sensitive species, then such stress should not be deemed a result of degradation or elimination of suitable habitat.	The systems should not be impacted by any point source or nonpoint discharges capable of preventing the maintenance of health biological communities after liming.	**Lakes** Lakes must have a retention time of ≥0.5 year, unless indirect liming technologies (e.g., stream or watershed liming) are used. **Streams** Streams must have a maximum mean monthly flow of ≤50 m^3/sec.	Lakes should have surface areas of ≥2 ha, unless they are drainage lakes and indirect liming technologies (e.g., stream or watershed liming) are used.

Maintenance liming	Systems must presently support biological populations or communities of special ecological importance (i.e., rare, threatened, or endangered aquatic or amphibious species requiring circumneutral water quality; fish populations of value as broodstock; special natural heritage strains of fish; or regionally unique species assemblages) or important recreational fisheries. If deleterious effects are indicated, such stress should not be attributable to nutrient enrichment, point source pollution, interspecific competition, or fishing pressures.	Surface water pH ≤6.5 or acid neutralizing capacity (ANC) values ≤50 μeq/L should exist in some significant spatial location in the system as determined sometime during the past 5 years. Such conditions should either (1) occur frequently, (2) be maintained for 2 or more weeks, or (3) occur during critical periods in sensitive life stages of important aquatic biota.	Same as mitigative liming	Same as mitigative liming	Same as mitigative liming	Same as mitigative liming
Research	May be waived	May be waived	May be waived	May be waived	May be waived	May be waived

[a] Criteria are identical for lakes and streams, except for the hydrology and size criteria.

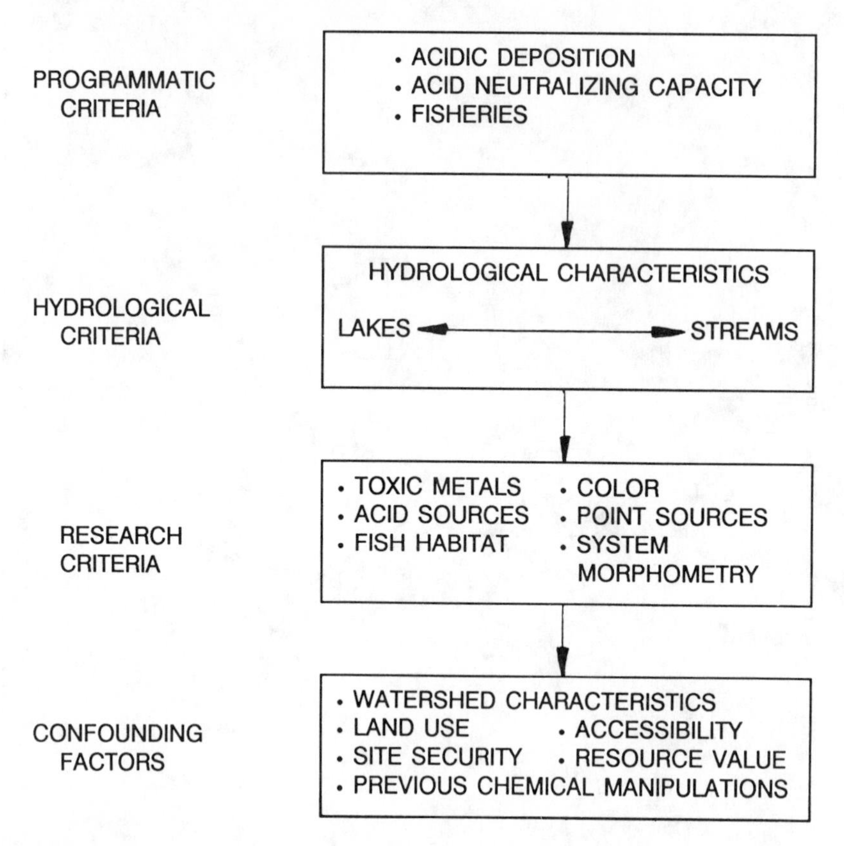

Figure 4.1 Factors considered for site selection in the U.S. Fish and Wildlife Service research program. (Source: Brown and Goodyear, 1987.)

μeq/L) in a significant spatial location of the water body at some time during the past 5 years. Acidic conditions should be evident that occur frequently and are sustained for 2 or more weeks or occur during critical periods for sensitive life stages of fish.

- The system must have an existing fish population or historical records indicating that fish were once present. If fish are present, stress should be indicated, as shown by missing year classes, loss of sensitive species, high incidence of morphological abnormalities, subnormal growth rates, poor condition factors, or poor recruitment. Stress should not be due to nutrient enrichment, point source pollution, or fishing pressure.
- Lakes should be drainage lakes with hydraulic retention times between 0.5 and 6.0 years, surface areas of at least 5.0

ha, and depths sufficient to allow distinguishable littoral and pelagic zones.

- Average annual stream discharge should exceed 0.2 m^3/sec, and the maximum monthly average should not exceed 25 m^3/sec. The treated segment of the study stream should be at least 3 km long.
- Concentrations of toxic or potentially toxic metals should be representative of surface waters with similar hydrology, and pH values in the region and should be derived only from natural watershed processes or atmospheric deposition.
- Aquatic chemistry should not be dominated by natural acidity or by acidic or basic point source discharges.
- Apparent color should not exceed 75 platinum cobalt units or dissolved organic carbon concentrations should not exceed 4.5 mg/L.
- Suitable habitat (spawning sites, nursery areas, shelter, and dissolved oxygen and temperature regimes) must be available for establishing or reestablishing regionally important fish species after liming.
- The drainage basin should be well defined and should not display highly variable soil development and subbasin differences in soils, vegetation, slope, etc. Recent natural or anthropogenic disturbances, such as fires or logging, should not be evident. Sites should be accessible throughout the year, yet reasonably protected from vandalism or other activities that may jeopardize field research efforts.
- The resource to be protected or restored should be economically, culturally, or ecologically important to the region.
- Chemical manipulations should not have occurred at the site within the last 10 years.

These site selection criteria were applied in the program, and by 1986 four sites were selected for treatment by direct (surface water) liming: a lake in Minnesota and three streams—one each in Massachusetts, West Virginia, and Tennessee.

Thrush Lake in Cook County, Minnesota is typical of acid-sensitive lakes in the upper midwestern United States (Schreiber, 1988). It supports a put-grow-and-take trout fishery, for lake trout and brook trout. In recent years both lake trout and brook trout stockings have generally been unsuccessful. The objective of the mitigation program at Thrush Lake, therefore, is to improve water chemistry so that stocking is successful.

Whetstone Brook, Franklin County, Massachusetts, was reported in 1953 to be the best trout stream in the central highland region of Massachusetts (Schreiber, 1988). Thereafter, the abundance of both native brook trout and wild brown trout declined, and thus the site was selected for mitigation to restore these fisheries. Limestone treatment at this site should determine the effectiveness of water-powered doser technology in improving the trout fishery.

Dogway Fork, in the Monongahela National Forest of West Virginia, was reported to have had brook trout in the 1960s (Schreiber, 1988). Acidic conditions there now prevent the maintenance of a year-round resident trout population in the stream. Liming activities were undertaken in this stream to restore a resident trout population and to determine the effectiveness of rotary drum treatment technology.

Laurel Branch, in eastern Tennessee, currently supports a diverse rainbow trout and forage fish community (R.K. Schreiber, personal communication, 1990). Evidence suggests, however, that the trout population becomes stressed during storm episodes. Since the stream is located in a region where little buffering capacity is afforded by the watersheds, it is considered one of the most acid sensitive in the state. This site was selected for liming to test the effectiveness of maintenance liming with a water-powered doser to maintain the existing fishery and to reduce stress during storm events.

Massachusetts Division of Fisheries and Wildlife

The purpose of the Massachusetts mitigation program, described in an environmental impact report in 1984 (Massachusetts Division of Fisheries and Wildlife, 1984), is to restore fisheries in the critical aquatic habitats of Massachusetts. The program focused on restoring water quality of selected ponds to sustain stocked trout before 1983 and was expanded in that year to restore ponds to support other fisheries, such as smallmouth bass and chain pickerel. For example, a number of ponds on Cape Cod were limed to enable smallmouth bass to reproduce.

The following items explain the decision points identified in

Figure 4.2 for the site selection criteria identified for lakes under the Massachusetts program:

- The lake or pond must have a maximum alkalinity of 40 μeq/L.
- Ponds must have public access and a minimum surface area of 5 ha, unless unique circumstances justify including the site in the mitigation program.
- Dystrophic bogs are excluded.
- Pond retention times must exceed 1 year. Exceptions are made for liming of refuge areas for large lakes or for protecting fish spawning and nursery areas.
- In candidate water bodies, the concentrations of metals such as aluminum, copper, lead, mercury, and zinc must be low or nil. Aluminum concentrations must not exceed 200 μg/L. No specific upper limits are provided for other metals.
- Lake bottoms must not be extensively covered with mud and organic matter that would bind a considerable portion of the applied limestone, resulting in rapid neutralization and oxidation of organic matter, followed by oxygen depletion.
- Limestone applications must be limited to about 1 tonne/ha applied primarily within the 5- to 7-m contour zone. When the mean depth of the pond or lake is less than or equal to 2.5 m or the maximum depth does not exceed 6 m, the dosage must not exceed 0.5 tonnes/ha.
- Ponds with high accumulations of organic material can be considered as experimental sites, and can be treated with small successive applications of limestone over several years.
- All candidate waters must be screened for the presence of rare and endangered species. If present, the recommendations of biologists will determine whether treatment is appropriate.
- The most cost-effective methods must be applied.

Liming of several sites, selected on the basis of these criteria, has allowed sport fish species to be stocked. Before Wallum Lake was limed, for example, only adult brook trout could be stocked; after liming, a trophy brown trout fishery was established and was supported by a forage population of alewives. No specific criteria were described for stream liming site selection.

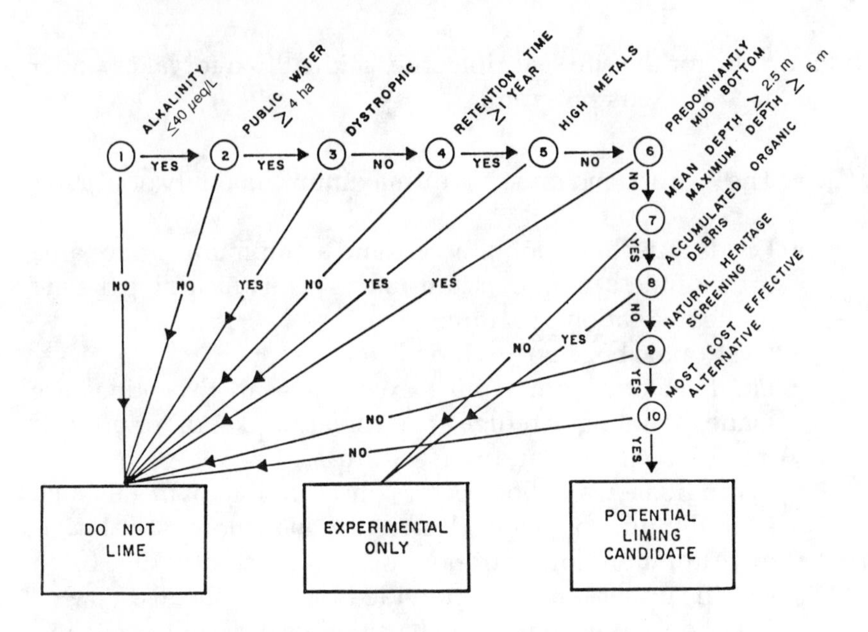

Figure 4.2 Decision chart for site selection under the Massachusetts operational liming program. (Source: Massachusetts Division of Fisheries and Wildlife, 1984.)

New York Department of Environmental Conservation

The New York Department of Environmental Conservation operational liming program focuses on the restoration of fisheries. The objective is to create water quality favorable for the survival of fish and other aquatic organisms by neutralizing lake and pond acidity. In 1988, the department prepared an environmental impact statement (Simonin et al., 1988) that contained revised policy guidelines for selecting candidate liming sites. A significant change in the guidelines from those adopted in 1984 was an increase in the criterion for the maximum pH from 5.5 to 5.7, enabling several additional candidate sites to be acceptable for inclusion in the program.

Site selection criteria for the revised program are described in more detail below.

- Summer surface water pH must be 5.7 or less, or the ANC must be 20 μeq/L or less.

- Inlets or outlets must be present, or floating sphagnum mats must occupy less than 25% of the lake basin area.
- Naturally acidic waters are excluded.
- Lake retention time must exceed 0.5 year.
- Dissolved oxygen and temperature levels during late summer and late winter must be suitable for the survival of the fish species considered for management, or become suitable after liming treatments.
- A lake must meet at least one of the following criteria: records should show a serious decline in a unique or previous fishery, a formerly heavy angling pressure due to its location, or represent a brood stock water that contains heritage strains of fish or populations of threatened or endangered fish species.

Ponds limed in the past that did not meet the above criteria remain as part of the program if they continue to respond favorably to periodic liming treatments. Other sites not meeting the criteria are included on a case-by-case basis with sufficient justification.

The operational focus and fisheries management aspects of this program are similar in many respects to those of the Massachusetts program. For example, maintaining natural populations of brook trout is a high priority of the department's fisheries management plan. Most lakes treated are in the Adirondacks, and several acidified waters, once devoid of fish, have been restored to support productive fisheries. However, it was estimated that 75% of the acidic waters in the Adirondacks are unsuitable for conventional whole lake liming because their retention times are less than 0.5 year.

Sweden

Like the U.S. mitigation programs, the Swedish operational liming program also focuses on restoring healthy fisheries in lakes and streams acidified by acidic deposition (Nyberg and Thornelof, 1988). The focus of the Swedish liming program has broadened somewhat in recent years to also consider all the natural flora and fauna. The program's goal is to neutralize all of Sweden's estimated 16,000 acidified lakes and 90,000 km of acidified streams. Inasmuch as government grants subsidize 85% of the liming costs, selection criteria for candidates are not particularly restrictive. All waters with pH less than 6.0 or alkalinity less than 50 μeq/L are eligible for inclusion. To date, about 5000 lakes and several streams have been treated (Nyberg, 1989).

Norway

The Norwegian operational liming program is similar to that of Sweden, with a policy to improve the sport fishing in acidified areas and to restore resident fish populations (Hindar and Rosseland, 1988; Langvatn, 1989). The goal of the program is to treat by 1997 an area corresponding to 20% of the area of the country that was acidified in 1980. To receiving support from the Norwegian government for liming the following conditions must be met:

- availability of fishing licenses to the public,
- local interest in the project and a plan for the liming operation,
- a cost-benefit evaluation, and
- water chemistry data indicating the need for liming.

Generally, waters with pH less than 6.0 or alkalinity less than 50 μeq/L are eligible for inclusion. To date, several hundred lakes and several large rivers have been treated (B.O. Rosseland, personal communication, 1990).

Other Considerations for Site Selection

Additional technical, physical, logistical, and societal factors specific to selected sites can influence the probability of successful treatment. Examples of factors that may complicate the treatment process at a site follow:

- presence of nontarget endangered species of fish, amphibians, or plants;

- slow or improbable issuance of permits required by public agencies;
- strong, adverse public sentiment regarding proposed management plans for the system;
- very large and complex aquatic systems; and
- absence of specific support services and facilities required for field operations.

Treatment Strategies Adopted for Major Mitigation Programs

Many methods are available for the mitigation of acidic surface waters by using liming treatments. Although research programs may test several different methods, practical operational programs cannot use unproven or experimental techniques. Next are described the technologies currently listed as acceptable for inclusion in major operational mitigation programs (summarized in Table 4.3).

Living Lakes, Inc.

The LLI program currently restricts the treatment strategy to the use of certain types of limestones and the application techniques described below (LLI, 1987).

Neutralizing Agents

The operations manual for the LLI program specifies limestone as the only neutralizing agent currently acceptable for use in liming treatments (LLI, 1987). Moreover, limestone slurry is recommended in most instances because of its greater dissolution efficiency and more precise and consistent distribution.

Purchasing limestone as a dry powder and producing the slurry on the site is considered an option. The use of limestone as a dry powder is also acceptable, on the basis of consideration of several factors that include the type of limestone and equipment available, and the relative costs.

Limestone should contain less than 5% $MgCO_3$. For lakes, the preferred limestone particle size has an average diameter be-

Table 4.3 Treatment Strategies Recommended for Major Mitigation Programs[a]

	Neutralizing Agents		Application Techniques: Lakes				Application Techniques: Streams			Application Techniques: Watersheds	
Program	Limestone	Other Materials	Boat	Truck	Aircraft	Doser	Diversion Well	Rotary Drum	Limestone Barrier	Any Proven Watershed Methods	Reference
Living Lakes, Inc.	Yes	No	Yes	No	Yes	Yes	Yes	No	No	Yes	LLI, 1987
U.S. Fish and Wildlife Service	Yes	No	Yes	No	Yes	Yes	Yes	Yes	No	No	Brown and Goodyear, 1987
Massachusetts	Yes	No	—[b]	—[b]	—[b]	—[b]	—[b]	—[b]	—[b]	—[b]	Massachusetts Divison of Fisheries and Wildlife, 1984
New York	Yes	Acceptable	—[b]	—[b]	—[b]	—[b]	—[b]	—[b]	—[b]	No	Simonin et al., 1988
Sweden	Yes	Acceptable	Yes	Yes	Yes	Yes	Yes	No	No	Yes	Nyberg and Thornelof, 1988
Norway	Yes	Acceptable	Yes	Yes	Yes	Yes	Yes	No	No	Yes	B.O. Rosseland, personal communication, 1990

[a] Yes = recommended.
No = not recommended.
Acceptable = may be allowed.

[b] No specific recommendations provided.

tween 6 and 20 μm. Particle sizes exceeding this range can be used for lakes with depths greater than 7–8 m or pH values below 5.4. For streams, the recommended particle size range is 3–10 μm for turbulent waters with pH values ≤5.4. Finer grades of limestone (0.5–5 μm) are recommended for slow-moving streams with pH values >5.4.

Application Techniques

For lake liming, application by boat or barge and aerial application by helicopter are recommended. Although helicopter application is costly, it is highly efficient and is the most suitable technique for treating remote lakes. During the first year of the LLI program, helicopters were used exclusively to treat 15 lakes (Brocksen and Emler, 1988). Boat and barge methods are less expensive than the helicopter method, and in 1987 were used exclusively to treat seven additional lakes. Limestone was applied from a boat with an onboard slurry box or a barge with a pneumatic tank system and spray arms.

Of the stream liming technologies mentioned in Chapter 2, only the use of limestone barriers is considered nonviable in the LLI operations manual. Stream treatments included a one-time application (in 1986) of limestone powder in Yokum Run, West Virginia and the use of electrically powered wet slurry dosers (in 1987) in two streams in Maryland (Brocksen and Emler, 1988). Other treatment devices installed in 1988 and 1989 but not yet fully tested included water-powered dosers installed in Maryland and Pennsylvania streams and a diversion well in Pennsylvania.

Watershed liming was considered in the LLI Operations Manual to be a viable option for treating lakes and streams. In October 1989 limestone was applied by helicopter to the Woods Lake watershed in New York, in cooperation with the Electric Power Research Institute and others (Brocksen et al., 1988; Boutacoff, 1990; Olem et al., 1990).

U.S. Fish and Wildlife Service

The service restricts treatment strategy to the use of various grinding grades of limestones and the application techniques described below (Fraser et al., 1985).

Neutralizing Agents

Consistent with the LLI program, the service operations manual states that limestone is the only neutralizing agent to be used in liming treatments. The only other requirement is that the limestone not exceed specified maximum levels of contaminants.

Application Techniques

Certain treatment methods considered acceptable for operational liming programs are not acceptable for use in the Service program. Application techniques are limited to those that are designed to meet short-term water quality targets, allow the quantities and dissolution rates of limestone to be controlled, and are applicable to a major subset of U.S. surface waters vulnerable to acidification.

Acceptable application techniques for lakes are fixed-wing aircraft or helicopter techniques and boat or barge methods. Acceptable stream liming techniques are electrically powered dry-powder and wet-slurry dosers; battery- and water-powered dosers; rotary drums; and fluidized diversion wells. Watershed liming is not considered an alternative application method for lake and stream treatment under this program, but is being investigated separately as a treatment strategy.

Thrush Lake was limed by boat; dry-powdered limestone and a slurry box were used (R.K. Schreiber, personal communication, 1990). Water-powered doser technology is being used for the Massachusetts and Tennessee stream sites, and the self-feeding, rotary drum system was installed in 1988 for the West Virginia stream site.

New York Department of Environmental Conservation

The New York operational liming program did not set strict requirements for treatment approaches; instead, recommendations based on information in the technical literature were made in the 1988 environmental impact statement (Simonin et al., 1988).

Neutralizing Agents

Two neutralizing materials, limestone and hydrated lime, have been used primarily in past projects in New York. The use of hydrated lime was discontinued in the 1970s, however, because its rapid dissolution caused excessive pH increases in some treatments, and its causticity made handling difficult. Agricultural limestone is considered the material of choice for most projects, but the use of alternative neutralizing materials is not precluded. For example, Bone Pond, an acidic surface water in Franklin County, was treated with sodium carbonate in 1987. Particle size distributions and limestone chemical composition are also to be considered.

Application Techniques

No specific requirements or guidelines for application techniques are described for the New York state liming program. Because the program focuses on treating lakes, many of which are remote, aerial application is described as an effective application technique. Other techniques include boat and barge methods and application with a truck or tractor on ice-covered lakes.

Watershed liming techniques to treat lakes and streams are not recommended, as they are viewed by the department as being in the experimental stage and not yet appropriate for an operational program.

Massachusetts Division of Fisheries and Wildlife

The Massachusetts environmental impact report provides only general guidance on neutralizing agents and application techniques to be used for the state program (Massachusetts Division of Fisheries and Wildlife, 1984). Specific requirements are not included, but ground agricultural limestone is recommended. Requirements for particle size and chemical composition are not detailed in the environmental report. No specific application techniques are recommended, although boat techniques and truck and tractor applications to ice-covered lakes have been used in past projects (Bergin, 1984).

Sweden

Requirements for liming lakes, streams, and watersheds under the Swedish program are purposely broad to allow counties desiring government assistance to make a specific proposal (Nyberg and Thornelof, 1988). All proposals are reviewed by the National Environmental Protection Board in cooperation with the National Board of Fisheries.

Neutralizing Agents

Many neutralizing agents were tested during the trial period of the operational program from 1977 to 1982 (Nyberg and Thornelof, 1988). The current program (1982 to the present) excludes practically all neutralizing agents except limestone. About 80% of the limestone used is finely ground (commercial particle size classification 0–0.2 mm); the other 20% is primarily a coarser product (0–0.5 mm). Maximum acceptable concentrations of metals and other contaminants in limestone to be used in the program are specified.

Application Techniques

Application methods are not restricted, but only a few methods have thus far been used. Lake methods used to date include application by boat, truck or tractor on ice, and helicopter. Direct stream application techniques have been limited to doser equipment and diversion wells. Watershed methods have consisted primarily of helicopter, truck, or tractor application to wetlands and seepage areas.

Norway

Requirements for liming surface waters and watersheds under

the Norwegian program are similar to those of Sweden (B.O. Rosseland, personal communication, 1990). These requirements are purposely broad to allow local fisherman associations, groups of landowners, and municipalities desiring government assistance to make a specific proposal (Hindar and Rosseland, 1988). All proposals are reviewed by the National Directorate for Nature Management. Neutralizing agents and application methods are not specifically restricted, but only limestone and a few application methods have thus far been used.

Target Water Quality

The principal water quality objective in restoration liming is to restore preacidification conditions in lake or stream systems so that sufficient ANC for fish survival and growth is provided. Although target pH and ANC may vary on the basis of the specific objective of the project, most programs to date have been concerned with fisheries management. Thus, the appropriate target water quality requirements are those suitable for a particular fish species and the fisheries management objectives.

The Fish and Wildlife Service program set target water quality levels for lakes on the basis of two criteria (Fraser et al., 1985). First, the target water chemistry was established to ensure that species diversity and standing crop of biota would be similar to those at the site before acidification and would be characteristic of lakes in the same geographic region that are not affected by acidic deposition. Second, the water chemistry conditions were to be maintained as long as possible before reliming was necessary. Treated lakes eventually reacidify if acidic inputs continue. Thus, water quality targets and the liming strategy must provide a balance between maintaining the circumneutral characteristics needed for biota of most softwater lakes and applying the largest possible dose to avoid the need for frequent reliming.

Because streams normally are treated continuously through the use of dosing devices, reacidification generally does not occur. For this reason, the Fish and Wildlife Service program set a lower target ANC level for streams than for lakes. Providing a higher level of ANC immediately downstream from the dosing device may be necessary, however, depending on acidic conditions in the target area, as when significant acidic tributary streams or seepage areas exist at some point downstream. In these instances, maintaining minimum ANC probably depends on spatial conditions in the streams that are roughly comparable to temporal conditions in lakes.

Target Levels for Major Mitigation Programs

Posttreatment water-column pH values are similar among the major mitigation programs (Table 4.4) and are set at levels between 6.0 and 6.5. All of the programs noted that these are minimum values and that target values for a particular lake or stream may vary. Target ANC concentrations were set only by the LLI, Fish and Wildlife Service, Sweden, and Norway. Minimum ANC concentrations were set at 100 μeq/L for lakes and for streams in the LLI, Swedish, and Norwegian liming programs. The ANC targets for the Fish and Wildlife Service program were set at 100 μeq/L for lakes and 50 μeq/L for streams. A lower target value for streams is sufficient because additional alkalinity (to prolong the time between one-time treatments) is not required when dosing is continuous.

Reliming Thresholds for Major Mitigation Programs

Lake reliming thresholds for pH or ANC also were set for most of the major mitigation programs to treat the surface water before reacidification occurs (see Table 4.4). ANC levels range between 10 and 50 μeq/L. All programs consider it mandatory that residual ANC be present between treatments. The reliming threshold values for ANC are generally the same as the maximum ANC criteria for site selection. Results of biological effects research generally indicate no acidification effects if ANC is maintained above these levels (Baker et al., 1990a). Other factors, such as maintaining calcium concentrations above 0.5 mg/L, may be as important as pH and ANC. Calcium has been shown to reduce the toxic effects of aluminum (Brown, 1982a,b). Fortunately, when limestone or other calcium-based products are applied, calcium often remains above 0.5 mg/L if pH or ANC are maintained at desired levels (Driscoll et al., 1989a).

Table 4.4 Minimum Water Quality Targets and Reliming Threshold Values Set by Major Mitigation Programs

	Posttreatment		Reliming Threshold		
Program	Minimum pH	Minimum ANC (μeq/L)	pH	ANC (μeq/L)	Reference
Living Lakes, Inc.	6.5	100[a]	—	50	LLI, 1987
U.S. Fish and Wildlife Service	6.5	Lakes—100 Streams—50	—	10	Fraser et al., 1985
Massachusetts	6.0 to 6.5	—	—	—	Bergin, 1984
New York	6.5[b]	—	6.0	25	Simonin, 1988
Sweden	6.0	100	6.0	50	Nyberg and Thornelof, 1988
Norway	6.5	100	6.0	50	B.O. Rosseland, personal communication, 1990

[a] Although 100 μeq/L was a minimum value for ANC, targets higher than 200 μeq/L generally were not recommended.
[b] Minimum target pH for brook trout; general target was set to be at least one pH unit higher than preliming value.

Reliming thresholds are selected to ensure that the frequency of sampling during posttreatment monitoring is sufficient to identify approximately the time when the limits will be reached. Specifying thresholds also ensures that a total loss of alkalinity will not occur before retreatment.

Estimates of Systems That May Benefit from Liming

Estimates of the number and percentage of lakes in the United States, Canada, Sweden, and Norway that may benefit from liming treatments are provided in this section (Table 4.5). The total surface area of acidic lakes and the total length of acidic streams are also estimated. These estimates were obtained from a review of studies designed to determine the location and extent of acidic surface waters.

These results should be used with extreme caution. They do not refer to the number of surface waters that are candidates for liming, but to the surface waters where pH or alkalinity are likely to be too low to support a healthy fishery. There are many other considerations, such as water retention time, surface area, stream flow, pollutant discharges, and accessibility. There should also be local support for the liming. For example, liming may be considered at a lake located near an area of the Adirondacks where fishing is popular, but would probably not be considered at a remote lake in Canada with similar water quality.

Also, there is considerable controversy over the accuracy of some of the figures. Different methods were used to determine these estimates, and the characteristics of the data base used in each country were different. For example, in the United States the National Surface Water Survey was specifically designed to provide population estimates, but Canadian estimates were determined on the basis of a number of studies of different design. Another factor not incorporated in the analysis is the presence of suitable physical habitat. The biota of many surface waters included in the figures may not benefit from liming because a suitable habitat does not exist.

The criteria used were different for each country because consistent information was not available. The figures are provided to allow general comparison of the surface waters in the United States that may benefit from mitigation with those in other

Table 4.5 Estimated Number of Surface Waters in the United States, Canada, Sweden, and Norway that May Benefit from Liming[a]

	Lakes				Streams	
Country	Total Number	Number Acidic	Percent of Total	Acidic Area (ha x 10^6)	Acidic Length (km)	Reference
United States[b]	43,944	2,525	9	—	36,000	Baker et al., 1990b
Canada[c]	—	14,000	—	—	—	Federal Provincial Research and Monitoring Coordinating Committee, 1988
Sweden[d]	85,000	16,000	18	0.5	90,000	Nyberg and Thornelof, 1988; Dickson, 1985; Berners and Thornelof, in press
Norway[b]	300,000	33,000	11	3.6	90,000	S. Sandoy, personal communication, 1990 (lakes); H.U. Sverdrup, personal communication, 1990 (streams)

[a] These results should be used with extreme caution. They do not refer to the number of surface waters that are candidates for liming, but to the surface waters where pH or alkalinity are likely too low to support a healthy fishery. Also, there is considerable controversy over the accuracy of some of the figures. Different methods were used to determine these estimates and the characteristics of the data base used in each country were different.

[b] pH $\leq$6.0.

[c] pH $\leq$5.0.

[d] Number of systems believed to be acidified (pH < 5.0 or alkalinity < 50 μeq/L) due to acidic deposition.

countries. Relative to Canada, Sweden, and Norway, fewer U.S. surface waters appear suitable for mitigation. Many more surface waters in Canada, however, are located in remote areas and do not have the accessibility or local support for mitigation.

United States

Baker et al. (1990b) presented an overview of the results of the National Surface Water Survey. The number of low pH lakes and streams in the surveyed regions were estimated from readings for waters sampled in the survey. The criterion of pH $\leq$6.0 was used to represent surface waters that might benefit from neutralization.

It was estimated that 2525 lakes and 36,000 km of streams in the surveyed regions had a pH $\leq$6.0. It should be noted that these estimates may include many naturally acidic systems. They do not, however, include lakes and streams known to be affected by acid mine drainage. Also, it is reasonable to expect many more surface waters with low pH values outside the surveyed regions. For example, Baker et al. (1990b) noted that acidic streams are found in parts of the Northeast just outside the Poconos-Catskills subregion.

Canada

The Federal Provincial Research and Monitoring Coordinating Committee (1988) estimated that there are currently 14,000 lakes in eastern Canada with a pH $\leq$5.0. Information was not provided on the source of the estimate nor on estimates of the total number of lakes in the region. Also, estimates were not provided for flowing waters. The results suggest, however, that there are proportionally far more low pH surface waters in Canada than in the United States.

Sweden

A 1985 survey of the water quality of 6908 lakes in Sweden led to the estimate that 21,500 of the 85,000 lakes with a surface area above 1 ha are acidic (Bernes and Thornelof, in press). Of these acidic lakes, about 16,000 are believed to have been affected by acidic deposition. The rest are considered to be naturally acidic. The estimates were based on lakes with pH <5 or alkalinity <50 μeq/L. The results of an application of the Henriksen model were used to distinguish between naturally acidic lakes and lakes acidified by acidic deposition (Bernes and Thornelof, in press). Nyberg and Thornelof (1988) estimated that the 16,000 lakes make up a total lake surface area of 0.5 million ha that meet the criteria for liming.

For streams, Dickson (1985) estimated that in Sweden approximately 90,000 km were affected by acidification. No information was reported on how the estimates for streams were obtained or the criteria used to define *acidic*.

It is apparent that there are many more low pH lakes and streams in Sweden compared to the United States. The total number of Swedish lakes reported was also higher than the number of lakes reported in the United States.

Norway

S. Sandoy (personal communication, 1990) reviewed data for lakes in Norway above 4 ha and estimated that about 33,000 waters are acidic and have a total surface area of 3.6 million ha. H.U. Sverdrup (personal communication, 1990) estimated that about 90,000 km of streams in Norway are acidic. According to these estimates, Norway also has many more low pH lakes and streams than the United States.

5

Summary and Conclusions

"Even if you are on the right track, you'll get run over if you just sit there."
Will Rogers

INTRODUCTION

Acidic conditions in surface waters can be mitigated by the addition of alkaline materials or by other methods in the lake, stream, or watershed. The primary objective of these strategies is the maintenance of water quality conditions suitable for the support of fish populations.

This report describes the techniques, costs, benefits, and overall effectiveness of mitigation strategies at the receptor (i.e., lake or stream). No comparisons of this strategy are made with mitigation strategies at the source (e.g., emission controls). The practice of mitigation of acidic surface waters will continue independent of emissions reductions; thus the techniques, effects, and effectiveness must be reviewed and assessed.

The objectives of this report are to

- summarize the various mitigation strategies that have been used or investigated to alleviate acidic conditions in surface waters,
- describe the physical, chemical, and biological effects of these strategies, and
- bring together and synthesize the technology (mitigation strategies) and the science (effects of mitigation) in one document.

These objectives are met through the use of all available literature on mitigation of acidic surface waters, including studies outside the United States.

MAJOR CONCLUSIONS

Conclusion 1

Liming can effectively mitigate many of the adverse ecological effects of surface water acidification independent of reductions in acidifying emissions.

Liming acidic surface waters and reducing emissions will continue independently because of several factors:

- Many surface waters will benefit only from liming because acidic inputs are from soils or other natural sources or from nonemission anthropogenic sources such as coal mining.
- Liming neutralizes acidic waters faster than emissions reductions.
- Although nearly all acidic surface waters could theoretically be mitigated by liming, there will be instances where it may not be considered economically feasible. Examples are lakes in remote locations and those so high in organic acids that extremely large doses are required.
- Some acidic waters, such as those in areas where intervention of any sort is prohibited, can be restored only by reducing acidic inputs.

Little experience exists for mitigation methods other than liming. Tentative conclusions on the feasibility of these other methods are given here.

- Nutrient addition alone has not proven effective in significantly increasing the acid neutralizing capacity of acidic surface waters.
- Introduction of acid-tolerant fish strains may improve the survival of stocked fish in slightly acidic waters. The effects of acidic conditions on fish are not mitigated, however, unless stocking is done in conjunction with neutralization.
- Pumping of water from a nearby source that contains alkalinity has been shown to be a viable technique for neutraliz-

ing acidic surface waters. It is not known whether the method has wide application or whether the costs of treatment compare favorably to liming.

Conclusion 2

Calcium carbonate ($CaCO_3$ or limestone) is the most effective overall neutralizing material for acidic surface waters.

Limestone is a natural component of surface water buffering systems, is easy to handle and noncaustic, can be obtained relatively free of contaminants, has a relatively slow dissolution rate that results in a gradual increase in pH, is widely available, and is relatively inexpensive. Limestone is the material most commonly used for treating surface waters today.

Many alkaline materials have been used or proposed for neutralization of acidic waters, including carbonates, oxides, hydroxides, and silicates of calcium, magnesium, and sometimes sodium. Use of base materials other than limestone are summarized below.

- Calcium hydroxide was commonly used before 1970, but it is rarely used today because of its causticity and the potential for excessive pH values if dosage rates are miscalculated.
- Sodium-containing materials, such as sodium carbonate, have been used in a few applications and have been recommended for injection into sediments. Sodium is more effective than calcium as an exchanger in cation exchange reactions. Sodium-based neutralizing agents, however, are higher in cost than those containing calcium and do not provide divalent cations, such as calcium, which have been shown to be beneficial for the growth and survival of aquatic organisms.

Conclusion 3

Conventional whole lake liming is a more established mitigation alternative than liming running waters and watersheds.

- Lakes have been the receptors most widely treated for mitigation of acidic conditions primarily because most can be treated with a single application that may last several years and the techniques are simpler than liming running waters. There are lakes and running waters that cannot be satisfactorily restored or protected from adverse ecological effects because of practical limitations in treatment methods and characteristics of the watershed and water body.
- Relatively few running waters have been treated to date, primarily because installation of permanent structures are generally required to provide continuous treatment and they experience technical problems. Their use has increased in recent years with the advances in automatic dosing devices, but they are not widely available commercially.
- Treatment of the watershed with limestone to protect lakes and streams has been receiving increased interest in recent years. Watershed liming has been shown to last longer than surface water liming and may provide increased protection from episodic acidic conditions and leaching of trace metals. Little experience exists for watershed liming; its use in the United States is experimental.

Reducing acidity of surface waters through liming was first reported in the early 1900s, primarily to enhance fish production in dystrophic systems or to neutralize acidity formed by pyrite oxidation as the result of coal mining. The general findings of a review of the history of liming are presented below.

- The first reported attempt to lime waters believed to be acidified by acidic deposition was conducted in Norway and Sweden in the early 1900s.
- Much of the early liming experience in the United States was conducted on waters receiving acid mine drainage.
- A few aquatic liming projects were conducted in the United States from the 1940s to 1960s on dilute acidic waters not exposed to mine drainage. These were primarily state programs (Massachusetts, Michigan, New York, West Virginia, and Wisconsin), operated on low budgets.
- More recent liming operations in the United States have been

conducted by the federal government (U.S. Environmental Protection Agency, U.S. Fish and Wildlife Service, U.S. Forest Service), state governments (Connecticut, Maryland, Massachusetts, Michigan, Minnesota, New York, Rhode Island, West Virginia, and Wisconsin), a private utility (Pennsylvania Power and Light), and not-for-profit organizations (Electric Power Research Institute and Living Lakes, Inc.).

- Liming experiments in the 1970s in Canada primarily involved the treatment of lakes affected by smelters near Sudbury, Ontario.

Conclusion 4

Several well proven, reliable, and commercially available liming technologies can produce predetermined target pH values and distribution patterns. Application methods vary according to the type of liming material used, the receptor (whether a surface water or its watershed), and the characteristics of the water body requiring treatment.

- Direct application of limestone slurry by boat is an effective method for lakes.
- Application by aircraft, particularly helicopters, is effective for lake and for watershed liming, particularly where access is limited.
- Land vehicles such as trucks and tractors have been used occasionally to transport and distribute liming materials. Their use is generally only effective for application to ice-covered lakes and to watersheds.
- Running waters are effectively treated with automated dosers for dispensing powdered or slurried limestone, but the devices are not widely available commercially. Diversion wells, rotary drums, and liming upstream lakes and the watershed are viable methods under special conditions. A limestone barrier is not an effective neutralization technique.
- Timing and location of application are important factors in the success of a liming operation. Application should precede (for lakes) or accommodate (for streams) critical events,

such as spring snowmelt or a sensitive period in fish life history (eggs, fry, spawning). Distribution over the entire surface of a lake is effective, but alternative placement in selected areas has sometimes proven beneficial.

Conclusion 5

Accurate methods and models are available for calculating the dose required to achieve ecological goals in the most cost effective manner, and for estimating the rates of reacidification of treated systems.

- Methods are available for planning single lake and stream applications, as well as multiple treatments to a system of lakes and streams. Some of the models are available on computer and give reliable predictions under a wide range of conditions. The most advanced methods use information on dissolution kinetics, chemical equilibria, and particle settling velocities to determine proper selection of material quantities and powder sizes.
- Flushing rate and pH (or acid neutralizing capacity) are the most important variables affecting model predictions.
- Simpler empirical methods provide similar results in most cases. Based on several years of data on required dosages, approximately 0.5–2 tonnes/ha/yr represents an average range, assuming direct application to a lake with a retention time of about 1 year.
- Models for calculating dose requirements for watersheds have recently been developed and have been applied operationally. Earlier watershed applications used general guidelines of 2–5 tonnes/ha to soils and 15–25 tonnes/ha to discharge areas.
- Several models are available for predicting the rate of reacidification of a lake treated with a one-time addition of base materials. The prediction has been found to be useful in determining whether a particular process or application is economically feasible. Because of the uncertainties of future precipitation patterns, reacidification models are less useful for determining when a particular lake must be retreated.

Conclusion 6

Liming surface waters commonly results in significant positive and predictable physicochemical changes in aquatic ecosystems.

- Physical changes that normally occur after addition of limestone to clear waters (color <50 units and dissolved organic carbon <5 mg/L) are decreased transparency and increased color and temperature. These changes often result from the suspension of undissolved neutralizing material or the increased biological activity that follows reduced toxicity of surface water after liming.
- Physical changes in colored waters (color >50 units and dissolved organic carbon >5 mg/L) include increased transparency and decreased color and temperature. These changes are often due to the precipitation of organic matter and chemical precipitation of particles that are less soluble at higher pH.
- The pH, ANC, dissolved inorganic carbon, and calcium of surface waters generally increase after limestone addition. There may be a period immediately after treatment—particularly with fine grade limestone—when pH is very high (pH 9.0), because dissolution exceeds the rate of CO_2 intrusion from the atmosphere.
- The concentrations of nutrients and organic matter may change after liming, but studies have not shown a consistent response in limed lakes. Phosphorus may increase slightly due to trace concentrations in the limestone or releases of phosphorus from the sediments. The dissolved organic carbon often decreases in colored waters due to the precipitation of humic substances and increases in clear waters when extensive mats of sphagnum are present.
- Metals that may be toxic to aquatic organisms—particularly aluminum, iron, lead, manganese, and zinc—are lower in limed waters due to precipitation, oxidation, surface adsorption, and ion exchange. Precipitation is often the dominant process due to reduced solubility of metal ions in neutral or alkaline waters.
- Limestone addition often causes changes in lake sediments,

due primarily to instantaneous adsorption of calcium. In many studies, the base neutralizing capacity decreased dramatically, selected metals increased, and changes occurred in other constituents such as organic matter, nutrients, and sulfate.

Conclusion 7

Liming generally increases nutrient cycling, decomposition, and primary productivity, and results in a positive response in fish and other aquatic biota.

The effects of liming on fish have received more attention than those on other aquatic biota because of the commercial, recreational, and resource management concerns related to fish survival, growth, and reproduction. The following conclusions can be drawn regarding the effects of liming on biological processes and biota.

- Nutrient cycling typically increases after liming primarily because of short-term phytoplankton dieback due to pH shock, release of phosphorus from sediments, and degradation of benthic sphagnum communities.
- Most aquatic systems respond to liming with increased rates of decomposition in sediments and increases in sediment bacterial numbers.
- Liming often promotes primary productivity in previously unproductive waters by release of phosphorus from sediments and the use of phosphorus content of the base material. The increases are primarily from increased production of phytoplankton.
- Liming often results in increased plankton biomass, but this is not observed in all limed lakes. There is an intricate relation between the response of phytoplankton, zooplankton, and fish in limed lakes.
- The community structure of benthic macroinvertebrates may be considerably altered as a result of liming, but there is no general relation between liming and the response of benthic fauna.

- The effects of liming have been favorable to fish populations. Liming has permitted the stocking of fish species previously lost from the system, introduction of new species, or the recovery of existing but stressed fish populations.
- Successful reproduction and growth of resident and reintroduced fish species have been observed in nearly all limed surface waters.
- In a few isolated instances, liming has caused mortalities in resident fish populations due to metal toxicity, but the conditions causing the toxicity were not always clearly identified.

Conclusion 8

To date, liming coupled with follow-up monitoring has been reported on about 150 lakes and 10 streams in the United States, over 5000 lakes and 100 streams in Sweden, several hundred waters in Norway, and a few waters in Canada, the United Kingdom, and other countries. There are fewer acidic surface waters in the United States than in Canada, Sweden, or Norway.

- Specific criteria for selecting surface waters suitable for liming were developed by six major mitigation programs. Candidate lakes generally have low pH and acid neutralizing capacity, a retention time over 0.5 year, suitable fish habitat, and a stressed fish population or absence of sport fish.
- Most liming programs include sites with pH $\leq$6.0 or 6.5 or ANC $\leq$50 or 100 μeq/L and use similar criteria for determining when lakes require retreatment.
- There are about 2525 acidic lakes and 36,000 km of acidic streams in the United States. A large fraction of these waters may benefit from mitigation.

SYNTHESIS

Acidic conditions in surface waters can be mitigated by adding alkaline materials to the lake, stream, or watershed or by other less common methods. The primary objective is the maintenance of water quality suitable for the support of fish populations. The mitigation strategy most effective for mitigation of acidic conditions is the addition of limestone.

Conventional whole-lake liming is a more established mitigation alternative than liming running waters or watersheds. Lakes have been the receptors most widely treated, primarily because most can be treated with a single application that may last several years. Limestone applications from boats or helicopters are generally the most effective techniques. Relatively few running waters have been treated to date; permanent structures are generally required to provide continuous stream-water treatment. Treatment of the watershed to protect lakes and streams has been receiving increased interest in recent years. Watershed liming has been shown to last longer than surface water liming and may provide increased protection from episodic acidic conditions and leaching of trace metals. Little experience exists for watershed liming; its use in the United States is experimental. Accurate methods are available for determining limestone doses required for treating lakes, streams, and watersheds, and for estimating lake reacidification rates.

The addition of base materials to surface waters commonly results in significant positive physical, chemical, and biological changes in aquatic ecosystems. Physical changes that normally occur in low humic waters after liming are decreased transparency and increased color and temperature. The pH, ANC, dissolved inorganic carbon, and calcium of surface waters generally increase after limestone addition. Concentrations of nutrients and organic matter do not significantly change after liming, but some studies have shown a response in limed lakes. Metals that may be toxic to aquatic organisms—particularly aluminum, iron, lead, manganese, and zinc—are sometimes lower in limed waters due to precipitation, oxidation, surface adsorption, and ion exchange. Limestone addition often causes changes in lake sediments, due primarily to the instantaneous adsorption of calcium.

Liming generally increases nutrient cycling, decomposition, and primary productivity, and results in a positive response in aquatic biota. Liming often results in increased plankton biomass and a considerable alteration in the community structure of benthic macroinvertebrates. The effects of liming have clearly been favorable to fish populations. Liming has permitted the stocking of fish species previously lost from the system, introduction of new species, or the recovery of existing but stressed fish populations. Successful reproduction and growth of resident and reintroduced fish species have been developed in many limed surface waters. In a few isolated instances, liming has caused mortalities in resident fish populations due to metal toxicity, but the conditions causing the toxicity were not always clearly identified.

Restoration of water quality conditions to those believed to exist before acidification has not always resulted in restoration of the original biota. It may also be possible that liming cannot restore conditions exactly as they were before acidification. For example, whole-lake liming does not eliminate acidic episodes from influent streams or from littoral zones. Other factors also separate limed waters from their unacidified counterparts, including precipitated metals, undissolved base material, elevated calcium levels, and the possibility of reacidification between treatments.

6

Remaining Knowledge Gaps and Research Needs

"It does not take much strength to do things, but it requires great strength to decide on what to do."

Elbert Hubbard

MITIGATION STRATEGIES

The techniques for applying base materials to lakes are well studied, and most methods are suitable under a variety of conditions. Stream liming methods, however, have not been as fully tested. Furthermore, some stream liming methods have either consistently failed or have not been studied sufficiently to determine their effectiveness. For example, limestone barriers and the placement of limestone fines in streams continue to be used, even though published studies over the last 20 years show that they have not performed adequately. The data from these unpublished treatments should be thoroughly evaluated to determine whether these methods, if redesigned, could be effective.

EFFECTS OF LIMING ON WATER CHEMISTRY AND BIOTA

Conventional whole lake liming has been conducted for many years, and its overall effects on water chemistry and biota have been, for the most part, thoroughly studied and fairly well understood. In contrast, the effects of watershed liming are not as well researched. No quantitative information exists—particularly in the United States—on the impacts of watershed liming on northern hardwood forests, bryophytes, acid tolerant vegetation, or wetlands. It is important that watershed liming research, such as the Woods Lake study, address these concerns.

Although the effects of whole-lake liming are generally well known, uncertainties remain about other equally important impacts: the results of liming warm water lakes, how metal precipitation affects sediment, how liming affects decomposition of organic matter in sediments, and the long-term effects of liming on the biota.

Few studies have considered the effects of liming on lakes in the southeastern United States. Florida has the highest number and percentage (23%) of acidic lakes in all the acid-sensitive regions of the United States (Baker et al., 1990b). There is considerable uncertainty about these sources of acidity, but they appear to be predominantly seepage systems dominated by inorganic acids: 60% of the lakes have a pH <5 and <2 mg/L of dissolved organic carbon. Furthermore, Florida lakes do not receive sudden acidic inputs, such as those that occur during spring snowmelt in the North Temperate zones. A comprehensive study of the ecological effects of liming warm water lakes is needed.

Weatherly (1988) suggested that long-term monitoring may be necessary to evaluate the important changes in biota after liming. For example, more than 6 years of posttreatment monitoring may be needed to determine the stability of colonizing populations of benthic macroinvertebrates. It may also be important to monitor a reference (untreated) lake because of variations in weather and precipitation patterns before and after liming. It is likely that the effects of liming reported in some studies were the result of factors other than the addition of limestone.

RESOURCE MANAGEMENT

Although surface water quality studies have characterized the lakes and streams in acid-sensitive regions of the United States, no quantitative inventory has been made of waters that may be candidates for liming. For example, there are many acidic lakes in the Adirondacks, but most are relatively small drainage lakes that have short retention times and therefore experience rapid reacidification. Furthermore, many of these lakes are remote and are unsuitable for liming. To provide a more accurate estimate of the surface waters suitable for liming, the number of waters that could be candidates for liming should be evaluated on a state-by-state basis. Maryland completed a statewide inventory in 1989 and has developed a plan for liming high priority waters.

The costs of liming have been reported in several studies, but costs have not always comparable because of foreign currency exchange rates, differences in labor and material costs between regions, and the inclusion of extensive research and planning costs in some evaluations. As such, reported costs for demonstration liming programs (the source of most cost data) are probably higher than the unit costs for typical operational liming programs in the United States. Future operational liming studies should include a cost evaluation component.

References

"You can listen to what everybody says, but the fact remains that you've got to get out there and do the thing yourself."

Joan Sutherland

Abrahamsen, H., and D. Matzow. 1984. Use of lime slurry for deacidification of running water. *Verh. Internat. Verein. Limnol.* 22:1981–1985.

Adamski, J., and M. Michalski. 1975. Reclamation of acidified lakes—Middle and Lohi, Sudbury, Ontario. *Verh. Internat. Verein. Limnol.* 19:1971–1983.

Adams, T.B., and R.W. Brocksen. 1988. Dose-response relationships for the addition of limestone to lakes and ponds in the northeastern United States. *Water Air Soil Pollut.* 41:137–164.

Alenas, I. 1986. *Kalkningsprojektet Harskogen—1976–1986.* Institutet For Vatten-Och Luftvardsforskning (IVL) Rapport L86/201, Goteborg, Sweden (in Swedish).

Almer, B., W. Dickson, C. Ekstrom, and E. Hornstrom. 1978. Sulfur pollution and the aquatic ecosystem , in: J.O. Nriago, ed. *Sulfur in the Environment II, Ecological Impacts,* John Wiley, New York, N.Y., pp. 273–311.

Altshuller, A.P., and R.A. Linthurst, eds. 1984. The acidic deposition phenomenon and its effects: Critical assessment review papers. *Volume II. Effects Sciences.* EPA-600/883-016BF, U.S. Environmental Protection Agency, Washington, D.C.

Andersson, P., and H. Borg. 1988. Effects of liming on the distribution of cadmium in water, sediment, and organisms in a Swedish lake. *Can. J. Fish. Aquat. Sci.* 45:1154-1162.

Appelberg, M., E. Degerman, A. Johlander, and L. Karlsson. Liming increases the catches of Atlantic salmon on the west coast of Sweden. *Nordic J. Freshwater Res.*, in press.

Arce, R.G., and C.E. Boyd. 1975. Effects of agricultural limestone on water chemistry, phytoplankton productivity, and fish production in soft water ponds. *Trans. Am. Fish. Soc.* 104:308-312.

Arnold, D.E. 1981. An Experimental limestone flow-through device for maintenance of pH in poorly buffered streams. *Cooperative Fisheries Research Unit Report*, Pennsylvania State University, University Park, PA, 2 pp.

Arnold, D.E., W.D. Skinner, and D.E. Spotts. 1988. Evaluation of three experimental low-technology approaches to acid mitigation in headwater streams. *Water Air Soil Pollut.* 41:385-406.

Baalsrud, K. 1985. Liming of Acid Water. Final Report, The Norwegian Liming Project. Report No. FRS, Department of the Environment, Arendal, Norway, 177 pp.

Baker, J.P. 1981. Aluminum Toxicity to Fish as Related to Acid Precipitation and Adirondack Surface Water Quality. Ph.D. Thesis, Cornell University, Ithaca, N.Y.

Baker, J.P. 1984. Effects of acidification on plankton, in: A.P. Altshuller and R.A. Linthurst, eds. The Acidic Deposition Phenomenon and Its Effects: Critical Assessment Review Papers. *Volume II. Effects Sciences.* EPA-600/883-016BF, U.S. Environmental Protection Agency, Washington, D.C., pp. 5-44 to 5-73.

Baker, J.P., and C.L. Schofield. 1982. Aluminum toxicity to fish in acidic waters. *Water Air Soil Pollut.* 18:289-309.

Baker, J.P., D.P. Bernard, S.W. Christensen, and M.J. Sale. 1990a. Biological effects of changes in surface water acid-base chemistry. State of Science and Technology Report 13, in: National Acid Precipitation Assessment Program, Acidic Deposition: State of Science and Technology, Volume II, September 1990.

Baker, L., P.R. Kaufmann, A.T. Herlihy, and J.M. Eilers. 1990b. Current status of surface water acid-base chemistry. State of Science and Technology Report 9, in: National Acid Precipitation Assessment Program, Acidic Deposition: State of Science and Technology, Volume II, September 1990.

Baker, L.A., T.E. Perry, and P.L. Brezonik. 1985. Neutralization of acid precipitation in softwater lakes. *Lake Reserv. Manage.* 1:356-360.

Bengtsson, B., W. Dickson, and P. Nyberg. 1980. Liming acid lakes in Sweden. *Ambio* 9:34-36.

Bengtsson, B., P. Nyberg, C. Wendt, T. Ahl, W. Dickson, and G. Persson. 1979. The Liming of Lakes and Waterways, 1977–1979: Experience Acquired, Measures Adopted, and the Needs of Continued Efforts. Report No. 8, National Swedish Board of Fisheries. National Swedish Environmental Protection Board, Goteborg, Sweden.

Bergin, J. 1984. Acid rain mitigation in Massachusetts, in: R.J. Fares and J.D. Kinsman, eds. The Proceedings of the Lake Acidification Mitigation Workshop, September 13–15, 1983, Albany, N.Y. Report 1359-02-84-CR, General Research Corporation, McLean, Va., pp. 3-12 to 3-17.

Bernes, C., and E. Thornelof. Present and future acidification of Swedish lakes: Model calculations based on an extensive survey, in: J. Kamari, ed., *Environmental Impact Models to Assess Regional Acidification*, in press.

Bernhoff, B. 1979. Acidification and lake liming. Cementa AB, Malmo, Sweden, 107 pp.

Blake, L.M. 1981. Liming acid ponds in New York. *N.Y. Fish Game J.* 28(2):208-214.

Booth, G.M., J.G. Hamilton, and L.A. Molot. 1986. Liming in Ontario: Short-term biological and chemical changes. *Water Air Soil Pollut.* 31:709-720.

Booth, G.M., J.G. Hamilton, and L.A. Molot. 1987. Experimental lake neutralization studies in Ontario 1981–1986. Final Technical Report. Ontario Ministry of the Environment, Ottawa, Canada.

Boutacoff, D. 1990. Working with the watershed. *EPRI J.* 15(1):29-33.

Boxholmkonsult A.B. 1989. Product information. Boxholm, Sweden.

Boyd, C.E. 1974. Lime requirements of Alabama fish ponds. Auburn University Agricultural Experiment Station, Bulletin 459, Auburn, Al., 19 pp.

Boyd, C.E. 1976. Liming farm fish ponds. Auburn University Agricultural Experiment Station, Leaflet 91, Auburn, Al., 7 pp.

Boyd, C.E. 1982a. Liming fish ponds. *J. Soil Water Conserv.* 37:86-88.

Boyd, C.E. 1982b. *Water Quality Management for Pond Fish Culture.* Elsevier Scientific Publishing Company, New York, 318 pp.

Boyd C.E., and M.L. Cuenco. 1980. Refinements of the lime requirement procedure for fish ponds. *Aquaculture* 21:293-299.

Boyd, C.E., and E. Scarsbrook. 1974. Effects of agricultural limestone on phytoplankton communities of fish ponds. *Arch. Hydrobiol.* 74:336-349.

Bradt, P.T., J.L. Dudley, and J.A. Remenar, Jr. 1989. Volume I, Part I. Biology and chemistry of two acid-stressed lakes in the Pocono Mountains: Changes in limestone treated and reference lakes, in: Effects of Neutralization and Acidification in Pocono Mountain Lakes, Report to Pennsylvania Power and Light Company, Allentown, Pa., pp. 1-172.

Brettum, P., and H. Hindar. 1985. Effects of lime treatment on the biological system. Lime Treatment Against Acid Water. Norwegian Institute for Water Research, Oslo, Norway.

Brewer, P.G., and J. C. Goldman. 1976. Alkalinity changes generated by phytoplankton growth. *Limnol. Oceanogr.* 21:108-117.

Broberg, O. 1988. Delayed nutrient responses to the liming of Lake Gardsjon, Sweden. *Ambio* 17:22-27.

Broberg, O., and G. Persson. 1984. External budgets for phosphorus, nitrogen, and dissolved organic carbon for the acidified Lake Gardsjon. *Arch. Hydrobiol.* 99:160-175.

Brocksen, R.W. 1990. Personal communication. Description of Living Lakes, Inc. aquatic liming program—1986–1989. Living Lakes, Inc., Washington, D.C.

Brocksen, R.W., and P.W. Emler, Jr. 1988. Living Lakes: An aquatic liming and fish restoration demonstration program. *Water Air Soil Pollut.* 41:85-94.

Brocksen, R.W., H.W. Zoettl, D.B. Porcella, R.F. Huettl, K-H. Feger, and J. Wisniewski. 1988. Experimental liming of watersheds: An international cooperative between the United States and West Germany. *Water Air Soil Pollut.* 41:455-472.

Brown, D.J.A. 1982a. The influence of calcium on the survival of eggs and fry of brown trout at pH 4.5. *Bull. Environ. Contam. Toxicol.* 28:664-668.

Brown, D.J.A. 1982b. The effect of pH and calcium on fish and fisheries. *Water Air Soil Pollut.* 18:343-351.

Brown, D.J.A. 1983. Effect of calcium on the survival of brown trout (*Salmo trutta*) at low pH. *Bull Environ. Contam. Toxicol.* 30:582-587.

Brown, D.J.A. 1988. The Loch Fleet and other catchment liming programs. *Water Air Soil Pollut.* 41:409-416.

Brown, D.J.A., G.D. Howells, T.R.K. Dalziel, and B.R. Stewart. 1988. Loch Fleet—a research watershed liming project. *Water Air Soil Pollut.* 41:25-42.

Brown, J.M., and C.D. Goodyear. 1987. Acid Precipitation Mitigation Program: Research methods and protocols. U.S. Fish and Wildlife Service, National Ecology Center, Leetown, W.V., NEC-87:27.

Bukaveckas, P.A. 1988a. Effects of lake liming on phytoplankton growth in acidic Adirondack lakes. *Water Air Soil Pollut.* 41:223-240.

Bukaveckas, P.A. 1988b. Effects of calcite treatment on primary producers in acidified Adirondack lakes. I. Response of macrophyte communities. *Lake Reserv. Manage.* 4:107-113.

Bukaveckas, P.A. 1989. Effects of calcite treatment on primary producers in acidified Adirondack lakes. II. Short-term response by phytoplankton communities. *Can. J. Fish. Aquat. Sci.* 46:352-359.

Bukaveckas, P.A. Effects of whole-lake base addition on the limnology of an acidic Adirondack lake (Woods Lake, N.Y., USA). *Verh. Internat. Verein. Limnol,* in press.

Burns, J.C., J.S. Coy, D.J. Tervet, R. Harriman, B.R.S. Morrison, and C.P. Quine. 1984. The Loch Dee Project: A study of the ecological effects of acid precipitation and forest management on an upland catchment in southwest Scotland. *Fish. Manage.* 15:145-167.

Cementa Movab. 1989. Product information. Malmo, Sweden.

Church and Dwight, Inc. 1989. Product brochure, Arm and Hammer brand sodium bicarbonate. Church and Dwight, Inc., New York, N.Y. 4 pp.

Clymo, R.S. 1973. The growth of sphagnum: Some effects of environment. *J. Ecol.* 61:849-869.

Dalpra, C.M., ed. 1988. Stream liming: Aiding fish habitat restoration. *Potomac Basin Reporter*. 44:3.

Dalziel, T.R.K., M.V. Proctor, and A. Dickson. 1988. Hydrochemical budget calculations for parts of the Loch Fleet catchment before and after watershed liming. *Water Air Soil Pollut.* 41:417-434.

Davis, J.E. 1988. The use of simulation models as resource management tools for restoring acidified waters. *Water Air Soil Pollut.* 41:435-454.

Davis, J.E., and R.A. Goldstein. 1988. Simulated response of an acidic Adirondack lake watershed to various liming mitigation strategies. *Water Resour. Res.* 24:525-532.

Davison, W., and W.A. House. 1988. Neutralizing strategies for acid waters: Sodium and calcium products generate different acid neutralizing capacities. *Water Res.* 22:577-583.

Degerman, E. 1987. Humosa Sjoar—En litteratursammanstallning med intriktning pa fisk och forsurning (Humic Lakes—a literature survey with emphasis on fish and acidification). *Naturvardsverket Rapport 3415*, Solna, Sweden, 72 pp. (in Swedish with English summary).

Degerman, E., and P. Nyberg. 1989. Effekster av sjokalkning pa fiskbestand (long-term effects of liming on fish populations in lakes). Nr 5-1989, Fiskeristyrelsens Sotvattenslaboratorium (Institute of Freshwater Research of the Swedish National Board of Fisheries), Drottningholm, Sweden, 35 pp. (in Swedish with English summary).

DePinto, J.V., R.D. Scheffe, W.G. Booty, and T.C. Young. 1989. Predicting reacidification of calcite treated acid lakes. *Can. J. Fish. Aquat. Sci.* 46:323-332.

DePinto, J.V., R.D. Scheffe, T.C.Young, W.G. Booty, and J.R Rhea. 1987. Use of acid lake reacidification model (AlaRM) to assess impact of bottom sediments on calcium carbonate treated lakes. *Lake Reserv. Manage.* 3:421-429.

Dickson, W. 1979. Liming of lakes. *J. R. Acad. Agric. For., Supplement* 13: 28-48.

Dickson, W. 1983. Liming toxicity of aluminum to fish. *Vatten* 39:400-404.

Dickson, W. 1985. Liming in Sweden, in: Liming Acidic Waters: Environmental and Policy Concerns. Center for Environmental Information, Inc., Albany, N.Y., p. 29.

Dickson, W., ed. 1988. Liming of Lake Gardsjon: An acidified lake in SW Sweden. Report 3426, National Swedish Environmental Protection Board, Solna, Sweden. 327 pp.

Dillon, P., N. Yan, W. Scheider, and N. Conroy. 1979. Acidic lakes in Ontario, Canada: Characterization, extent and responses to base and nutrient additions. *Arch. Hydrobiol. Beih.* 13:317-336.

Dilworth, C. 1988. Future liming in Sweden. *Acid Mag.* 6:36.

Dodge, D.P., G.M. Booth, L.A. Richman, W. Keller, and F.D. Tomassini. 1988. An overview of lake neutralization experiments in Ontario. *Water Air Soil Pollut.* 41:75-84.

Driscoll, C.T., W.A. Ayling, J.F. Fordham, and L.O. Oliver. 1989a. The chemical response of $CaCO_3$ treated lakes to reacidification. *Can. J. Fish. Aquat. Sci.* 46:258-267.

Driscoll, C.T., J.P. Baker, J.J. Bisogni, and C.L. Schofield. 1980. Effect of aluminum speciation on fish in dilute acidified waters. *Nature (London)* 284:161-164.

Driscoll, C.T., J.F. Fordham, W.A. Ayling, and L.O. Oliver. 1989b. Short term changes in the trace metal chemistry of two acidic lakes following calcium carbonate treatment. *Can. J. Fish. Aquat. Sci.* 46:249-257.

Driscoll, C.T., J.R. White, C.G. Schafran, and J.D. Rendall. 1982. Calcium carbonate neutralization of acidified surface waters. *J. Environ. Eng. Div.* ASCE. 1128-1145.

Dudley, J.L., and P.T. Bradt. 1987. Chemical changes in White Deer Lake (Pennsylvania) before and after limestone application. *Lake Reserv. Manage.* 3:430-435.

Dutkowsky, D., and F.C. Menz. 1985. A cost function for neutralizing acidic Adirondack surface waters. *J. Environ. Manage.* 12:277-285.

Edman, G., and S. Fleischer. 1980. The River Hogvadsan liming project—a presentation, in: D. Drablos and A. Tollan, eds. Ecological Impact of Acid Precipitation: Proceedings of an International Conference, March 11–14, 1980, Sandefjord, Norway, pp. 300-301.

Elser, M.M., J.J. Elser, and S.R. Carpenter. 1986. Paul and Peter Lakes: A liming experiment revisited. *Am. Midl. Nat.* 116:282-295.

Eriksson, F. 1988. Makrofytvegetation I kalkade sjoar (Macrophyte vegetation in limed lakes). Nr 9-1988, Fiskeristyrelsens Sotvattenslaboratorium (Institute of Freshwater Research of the Swedish National Board of Fisheries), Drottningholm, Sweden, 25 pp. (in Swedish with English summary).

Eriksson, F., E. Hornstrom, P. Mossberg, and P. Nyberg. 1983. Ecological effects of lime treatment of acidified lakes and rivers in Sweden. *Hydrobiologia* 101:145-164.

Evans, R.A. 1989. Effects of liming and brook trout (Salvelinus fontinalis) introduction on limnetic insect populations of two acidified, fishless lakes. *Can. J. Fish. Aquat. Sci.* 46:295-305.

Fares, R.J., and J.D. Kinsman, eds. 1984. The proceedings of the Lake Acidification Mitigation Workshop, September 13–15,

1983, Albany, NY. Report 1359-02-84-CR, General Research Corporation, McLean, Va.

Federal Provincial Research and Monitoring Coordinating Committee. 1988. Assessment of the state of knowledge on the long-range transport of air pollutants. Part 3, aquatic effects. Federal Provincial Research and Monitoring Coordinating Committee, Ottawa, Canada.

Fisher, D.L., ed. 1984. Groundwater neutralization 'initial success' in restoring fish to acid-impacted streams. Institute for Research on Land and Water Resources, Newsletter 15:5-6.

Fisher, D.L., ed. 1985. Acidic stream section neutralized. Institute for Research on Land and Water Resources, Newsletter 16:1-3.

Fjellheim, A., and G.G. Raddum. 1988. Birch leaf processing and associated macroinvertebrates in an acidified lake subjected to liming. *Hydrobologia* 157:89-94.

Flick, W.A., C.L. Schofield, and D.A. Webster. 1982. Remedial action for intensive maintenance of fish stocks in acidified water, in: R.E. Johnson, ed. Acid Rain/Fisheries, American Fisheries Society, Bethesda, Md., pp. 287-294.

Fordham, G.F., and C.T. Driscoll. 1989. Short-term changes in the acid/base chemistry of two acidic lakes following calcium carbonate treatment. *Can. J. Fish. Aquat. Sci.* 46:306-314.

Fraser, J.E., and D.L. Britt. 1982. Liming of acidified waters: A review of methods and effects on aquatic ecosystems. FWS/OBS-80/40.13, U.S. Fish and Wildlife Service, Washington, D.C., 189 pp.

Fraser, J.E., D.L. Britt, J.D. Kinsman, J. DePinto, P. Rogers, H. Sverdrup, and P. Warvinge. 1985. APMP guidance manual—Volume II: Liming materials and methods. U.S. Fish and Wildlife Service, Biological Report 80(40.25), 197 pp.

Fraser, J.E., D. Hinckley, R. Burt, R. Rodensky Severn, and J. Wisniewski. 1982. Feasibility study to utilize liming as a technique to mitigate surface water acidification. EPRI EA-2362, Electric Power Research Institute, Palo Alto, Ca., 101 pp.

Fraser, J., J. Kinsman, D. Britt, and D. Minnick. 1984. Assessment of potential mitigative techniques that may be applied to receptors affected by acidic deposition: Phase III. Revised Draft Report C84-006. International Science & Technology, Inc., Reston, Va.

Gagen, C.J., W.E. Sharpe, D.R. DeWalle, and W.G. Kimmel. 1989. Pumping alkaline groundwater to restore a put-and-take trout fishery in a stream acidified by atmospheric deposition. *N. Am. J. Fish. Manage.* 9:92-100.

Gahnstrom, G. 1988. The effects of liming on sediment oxygen uptake in Lake Gardsjon, Sweden, in: W. Dickson, ed. Liming of Lake Gardsjon: An Acidified Lake in SW Sweden. Report 3426, National Swedish Environmental Protection Board, Solna, Sweden, pp. 207-244.

Gahnstrom, G., G. Anderson, and S. Fleischer. 1980. Decomposition and exchange processes in acidified lake sediment, in: D. Drablos and A. Tollan, eds. Ecological Impact of Acid Precipitation: Proceedings of an International Conference, March 11–14, 1980, Sandefjord, Norway, pp. 306-307.

Garrison, P. 1984. Liming in Wisconsin, in: R.J. Fares and J.D. Kinsman, eds. The Proceedings of the Lake Acidification Mitigation Workshop, September 13–15, 1983, Albany, N.Y. Report 1359-02-84-CR, General Research Corporation, McLean, Va., pp. 3-72 to 3-90.

Gedrem, T. 1980. Genetic variation in acid tolerance in brown trout, in: D. Drablos and A. Tollan, eds. Ecological Impact of Acid Precipitation: Proceedings of an International Conference, March 11–14, 1980, Sandefjord, Norway, pp. 308-309.

Gee, A.S., and J.H. Stoner. 1989. A review of the causes and effects of acidification of surface waters in Wales and potential mitigation techniques. *Arch. Environ. Contam. Toxicol.* 18:121-130.

Gencsoy, H.T., J.G. Pappajohn, G.A. Clites, and P.E. Zurbuch. 1983. Design for an economic and efficient treatment station for acidic streams, in: T.N. Veziroglu, ed. *Alternate Energy Sources V, Part E: Nuclear/Conservation/Environment*. Elsevier Science Publishers B.V., Amsterdam, Netherlands, pp. 371-380.

General Research Corporation. 1985. Lake acidification mitigation project (LAMP) second annual technical report, November 1984–September 1985 Report No. 1397-03-85-CR, General Research Corporation, McLean Va., Volume II, Appendices.

Gloss, S.P., C.L. Schofield, and M.D. Marcus. 1989a. Liming and fisheries guidelines for acidified lakes in the Adirondack region. U.S. Fish and Wildlife Service, Biological Report 80(40.27), 59 pp.

Gloss, S.P., C.L. Schofield, and R.E. Sherman. 1988a. An evaluation of New York State lake liming data and the application of models from Scandinavian lakes to Adirondack lakes. *Water Air Soil Pollut*. 41:241-278.

Gloss, S.P., C.L. Schofield, and R.E. Sherman. 1988b. An evaluation of New York State lake liming data and the application of models from Scandinavian lakes to Adirondack lakes. Progress Report, NERC-88/04, U.S. Fish and Wildlife Service, 41 pp.

Gloss, S.P., C.L. Schofield, R.L Spateholts, and B.A. Plonski. 1989b. Survival, growth reproduction and diet of brook trout (*Salvelinus fontinalis*) in lakes after liming. *Can. J. Fish. Aquat. Sci.* 46:277–286.

Goodchild, G., and J.G. Hamilton. 1983. Ontario's experimental neutralization program, in: Lake Restoration, Protection and Management. EPA 440/5-83-001, U.S. Environmental Protection Agency, Washington, D.C., pp. 92-95.

Grahn, O., and H. Hultberg. 1975. The neutralization capacity of 12 different lime products used for pH-adjustment of acid water. *Vatten* 2:120-132.

Grahn, O., and O. Sangfors. 1988. A comparative study of macrophytes in Lake Gardsjon, during acid and limed conditions, in: W. Dickson, ed. Liming of Lake Gardsjon: An Acidified Lake in SW Sweden. Report 3426, National Swedish Environmental Protection Board, Solna, Sweden, pp. 281-308.

Greening, H., A. Janicki, and W.P. Saunders. 1987. An evaluation of stream liming effects on water quality and yellow perch spawning in Maryland coastal plain streams. 1986 Final Report. Maryland Department of Natural Resources, Tidewater Administration, Annapolis, Md.

Greening, H.S., A.J. Janicki, R.J. Klauda, D.M. Baudler, D.M. Levin, and E.S. Perry. An evaluation of stream liming effects on water quality and anadromous fish spawning in Maryland coastal plain streams: 1988 results. Final Report to Living Lakes, Inc., Washington, D.C. and Maryland Department of Natural Resources, Annapolis, Md., in press.

Gunn, J.M., and W. Keller. 1980. Enhancement of the survival of rainbow trout (*Salmo gairdneri*) eggs and fry in an acid lake through incubation in limestone. *Can. J. Fish. Aquat. Sci.* 37:1522-1530.

Gunn, J.M., and W. Keller. 1981. Emergence and survival of lake trout (*Salvelinus namaycush*) and brook trout (*S. fontinalis*) from artificial substrates in an acid lake. Ontario Fisheries Technical Report Series, 1, Toronto, Ontario, Canada. 32 pp.

Gunn, J.M., and W. Keller. 1984. *In situ* manipulation of water chemistry using crushed limestone and observed effects on fish. *Fisheries (Bethesda)* 9(1):19-24.

Gunn, J.M., J.G. Hamilton, G.M. Booth, C.D. Wren, G.L. Beggs, H.J. Rietveld, and J.R. Munro. 1990. Survival, growth, reproduction of lake trout (*Salvelinus namaycush*) and yellow perch (*Perca flavescens*) after neutralization of an acidic lake near Sudbury, Ontario. *Can. J. Fish. Aquat. Sci.* 47:446-453.

Gunn, J.M., M.J. McMurtry, J.M. Casselman, W. Keller, and M.J. Powell. 1988. Changes in fish community of a limed lake near Sudbury, Ontario: Effects of chemical neutralization or reduced atmospheric deposition of acids? *Water Air Soil Pollut.* 41:113-136.

Haines, T.A. 1981. Acidic precipitation and its consequences for aquatic ecosystems: A review. *Trans. Am. Fish. Soc.* 110:669-706.

Haines, T.A., and C.L. Schofield. 1980. Responses of fish to acidification of streams and lakes in eastern North America, in: Restoration of Lake and Inland Waters: Proceedings of an International Symposium on Inland Waters and Lake Protection, EPA 440/5-81-010, U.S. Environmental Protection Agency, Washington, D.C., pp. 467-471.

Hall, R., G. Likens, S. Fiance, and G. Hendrey. 1980. Experimental acidification of a stream in the Hubbard Brook Experimental Forest, New Hampshire. *Ecology* 61:976-989.

Hasler, A.D., O.M. Brynildson, and W.T. Helm. 1951. Improving conditions for fish in brown-water bog lakes by alkalization. *J. Wildl. Manage.* 15:347-352.

Hasselrot, B., and H. Hultberg. 1984. Liming of acidified Swedish lakes and streams and its consequences for aquatic ecosystems. *Fisheries (Bethesda)* 9(1):4-9.

Henrikson, L. 1988. Effects on water quality and benthos of acid water inflow into limed Lake Gardsjon, in: W. Dickson, ed. Liming of Lake Gardsjon: An Acidified Lake in SW Sweden. Report 3426, National Swedish Environmental Protection Board, Solna, Sweden, pp. 309-327.

Henrikson, L., and H.G. Oscarson. 1984. Lime Influence on Macro-Invertebrate Zooplankton Predators. Report 61, Institute of Freshwater Research, National Swedish Board of Fisheries, Lund, Sweden, pp.93-103.

Henrikson, L., H.G. Oscarson, and J.A.E. Stenson. 1984. Development of the crustacean zooplankton community after lime treatment of the fishless Lake Gardsjon, Sweden. Report 61,

Institute of Freshwater Research, National Swedish Board of Fisheries, Lund, Sweden, pp. 104-114.

Herrick, C.N., ed. 1987. Interim Assessment—The Causes and Effects of Acidic Deposition, Volumes I–IV. National Acid Precipitation Assessment Program, Washington, D.C.

Hindar, A., and B.O. Rosseland. 1988. Liming acidic waters in Norway: National policy and research and development. *Water Air Soil Pollut.* 41:17-24.

Hornstrom, E. and C. Ekstrom. 1986. Acidification and liming effects on phyto- and zooplankton in some Swedish west coast lakes. Report 1864, Swedish Environmental Protection Board, Solna, Sweden. 108 pp.

Howells, G.D., ed. 1986. The Loch Fleet project: A report on the pre-intervention phase (1) 1984-1986. Publication CEGB, SSEB, NSHEB, BC.

Howells, G. 1987. Acidity mitigation in a small upland lake, in: H. Barth, ed. *Reversibility of Acidification*. Elsevier Applied Science Publishers, Ltd., Essex, England, pp.104-111.

Howells, G.D., and D.J.A. Brown. 1986. Loch Fleet: Techniques for acidity mitigation. *Water Air Soil Pollut.* 31:817-825.

Huckabee, J.W., J.S. Mattice, L.F. Pitelka, D.B. Porcella, and R.A. Goldstein. 1989. An assessment of the ecological effects of acidic deposition. *Arch. Environ. Contam. Toxicol.* 18:3-27.

Hultberg, H., and I.B. Andersson. 1982. Liming of acidified lakes induced long-term changes. *Water Air Soil Pollut.* 18:311-331.

Hultberg, H., and U. Nystrom. 1988. The role of hydrology in treatment duration and reacidification in the limed Lake Gardsjon, in: W. Dickson, ed. Liming of Lake Gardsjon: An Acidified Lake in SW Sweden. Report 3426, National Swedish Environmental Protection Board, Solna, Sweden, pp. 95-134.

Hultman, B., C. Forsberg, and R. Engvall. 1983. Can acidification of lakes and rivers be cured by treated sewage? *Vatten* 39:3-14.

Hunn, J.B., L. Cleveland, and D.R. Buckler. 1985. Additional calcium reduces effects of low pH and aluminum on developing brook trout eggs and fry. Research Information Bulletin, U.S. Fish and Wildlife Service, Columbia National Fisheries Research Laboratory, Columbia, Mo.

Jackson, M.B., E.M. Vandermeer, N. Lester, G.A. Booth, L.A. Molot, and I.M. Gray. 1990. Effects of neutralization and early reacidification on filamentous algae and macrophytes in Bowland Lake. *Can. J. Fish. Aquat. Sci.* 47:432-439.

Janicki, A., and H. Greening. 1988a. An evaluation of stream liming effects on water quality and anadromous fish spawning in Maryland coastal plain streams: 1987 results. Final Report, Power Plant Research Program, Maryland Department of Natural Resources, Annapolis, Md.

Janicki, A., and H.S. Greening. 1988b. The effects of stream liming on water chemistry and anadromous yellow perch spawning success in two Maryland Coastal Plain streams. *Water Air Soil Pollut.* 41:359-384.

Jeffries, D. C. Cox, and P. Dillon. 1979. Depression of pH in lakes and streams in central Ontario during snowmelt. *J. Fish. Res. Board Can.* 36:640-646.

Johannessen, M., and O.K. Skogheim. 1985. Hydrochemical effects of lime treatment, in: M. Johannessen and D. Matzow, eds. Lime Treatment Against Acid Water, Report No. FRS, Norwegian Institute for Water Research, Oslo, Norway, pp. 57-78.

Johannessen, M., A. Skartveit, and R. Wright. 1980. Streamwater chemistry before, during, and after snowmelt, in: D. Drablos and A. Tollan, eds. Ecological Impact of Acid Precipitation: Proceedings of an International Conference, March 11–14, 1980, Sandefjord, Norway, pp. 224-225.

Johansson, S., and T. Nilsson. 1988. Lake Gardsjon— the hydrological balance of an acidified forested catchment area, 1979–1986, in: W. Dickson, ed. Liming of Lake Gardsjon: An Acidified Lake in SW Sweden. Report 3426, National Swedish Environmental Protection Board, Solna, Sweden, pp. 9-36.

Johnson, W.E., and A.O. Hasler. 1954. Rainbow trout production in dystrophic lakes. *J. Wildl. Manage.* 18:113-134.

Knauer, D. 1990. Personal communication. Description of Wisconsin Department of Natural Resources research on the feasibility of pumping alkaline groundwater to seepage lakes in Wisconsin—1987–1989. Wisconsin Department of Natural Resources, Madison, Wisconsin.

Knauer, G. 1986. Liming of a final cut acid lake. *Lake Reserv. Manage.* 2:382-385.

Keller, W., J. Gunn, and N. Conroy. 1980. Acidification impacts on lakes in the Sudbury, Ontario, Canada area, in: D. Drablos and A. Tollan, eds. Ecological Impact of Acid Precipitation: Proceedings of an International Conference, March 11–14, 1980, Sandefjord, Norway, pp. 228-229.

Keller, W., D.P. Dodge, and G.M. Booth. 1990a. Experimental lake neutralization program—overview of neutralization studies in Ontario. *Can. J. Fish. Aquat. Sci.* 47:410-411.

Keller, W., L.A. Molot, R.W. Griffiths, and N.D. Yan. 1990b. Changes in the zoobenthos community of acidified Bowland Lake after whole-lake neutralization and lake trout (*Salvelinus namaycush*) reintroduction. *Can. J. Fish. Aquat. Sci.* 47:440-445.

Kitchell, J.A., and J.F. Kitchell. 1980. Size-selective predation light transmission and oxygen stratification: Evidence from the recent sediments of manipulated lakes. *Limnol. Oceanogr.* 25:389–402.

Kretser, W., and J. Colquhoun. 1984. Treatment of New York's Adirondack Lake by liming. *Fisheries* (Bethesda) 9(1):36-41.

Krogerstrom, L. 1988a. Lime doser saved fishing club's trout. *Acid Mag.* 7:13.

Krogerstrom, L. 1988b. Salmon spawning again in limed river. *Acid Mag.* 7:10-12.

Langvatn, A. 1989. Kalking i vann og vassdrag (Liming in waters and watercourses). Nr. 4-1989, Direktoratet for Naturforvaltning (Directorate for Nature Management), Trondheim, Norway, 92 pp. (in Norwegian).

Larsson, L.S. 1988. Liming effects on phytoplankton in Lake Gardsjon, 1982–1985, in: W. Dickson, ed. Liming of Lake Gardsjon: An Acidified Lake in SW Sweden. Report 3426, National Swedish Environmental Protection Board, Solna, Sweden, pp. 245-280.

Lazarek, S. 1986. Responses of the Lobella-Epiphyte complex to liming of an acidified lake. *Aquat. Bot.* 25:73-81.

Leedy, D.L., and J.M. Franklin. 1981. Coal surface mining reclamation and fish and wildlife relationships in the eastern United States. Report No. FWS/OBS-80/25, U.S. Fish and Wildlife Service, Eastern Energy and Land Use Team, Kearneysville, W. V., 169 pp.

Lepley, R.H. 1989. Personal communication. Description of commercially available barge for application of limestone slurry to lakes. Sweetwater Technology Corporation, Palmer, Pa.

Lessmark, O., and E. Thornelof. 1986. Liming in Sweden. *Water Air Soil Pollut.* 31:809.

Lessmark, O. 1987. Markkalkning som metod for att motverka forsurning av sjoar och vattendrag (Ground liming as a method of mitigating acidification of surface waters). Nr 9-1987, Fiskeristyrelsens Sotvattenslaboratorium (Institute of Freshwater Research of the Swedish National Board of Fisheries), Drottningholm, Sweden, 38 pp. (in Swedish with English summary).

Lewis, C., and R. Boynton. 1976. Acid neutralization with lime for environmental control and manufacturing processes. Bulletin No. 216, National Lime Association, Washington, D.C., 16 pp.

Lindmark, G.K. 1982. Acidified lakes: Sediment treatment with sodium carbonate—a remedy? *Hydrobiologia* 92:537-547.

Lindmark, G.K. 1985. Sodium carbonate injected into sediment of acidified lakes: A case study of Lake Lilla Galtsjon treated in 1980. *Lake Reserv. Manage.* 1:89-93.

Living Lakes, Inc. 1987. Field operations program implementation manual for aquatic liming and fish restoration demonstration program. Living Lakes, Inc., Washington D.C.

Lovell, H. 1973. An appraisal of neutralization processes to treat coal mine drainage. EPA-670/2-73-093, U.S. Environmental Protection Agency, Washington, D.C., 81 pp.

Majumdar, S.K., O. Delucia, R.W. Holt, C.C. Derivaux, R.J. Marcus, T.J. Irwin, L.C. Mineo, P.M. Steed, and R.L. Morris. 1987. Effects of limestone treatment on microbial communities and sediment chemistry in acid-stressed Pennsylvania lakes, U.S.A, in: R. Perry, R.M. Harrison, J.N.B. Bell, and J.N. Lester, eds. *Acid Rain Scientific and Technical Advances*, Selper Ltd., pp. 465-485.

Majumdar, S.K., O. Delucia, C.C. Derivaux, P. Steed, T.J. Irwin, R.W. Holt, and S.W. Cline. 1989. Bacteriological studies of two acid-stressed Pocono lakes before and after limestone application, in: Effects of Neutralization and Acidification in Pocono Mountain Lakes, Report to Pennsylvania Power and Light Company, Allentown, Pa., pp. II-1 to II-70.

Marcus, M.D. 1988. Differences in pre- and post-treatment water qualities for twenty limed lakes. *Water Air Soil Pollut.* 41:279-292.

Massachusetts Division of Fisheries and Wildlife. 1984. Statewide liming of acidified waters. Generic Environmental Impact Report, Massachusetts Division of Fisheries and Wildlife, Westboro, Ma.

Matzow, D., B.O. Rosseland, and O.K. Skogheim. 1985. Effects of lime treatment on fish, in: M. Johannessen and D. Matzow, eds. Lime Treatment Against Acid Water, Report No. FRS, Norwegian Institute for Water Research, Oslo, Norway, pp. 109–128.

Melander, T. 1988. Saving Lakes. Informator AB, Gothenburg, Sweden, 115 pp.

Menz, F.C., and C.T. Driscoll. 1983. An estimate of the costs of liming to neutralize acidic Adirondack surface waters. *Water Resour. Res.* 19:1139-1149.

Molot, L.A. 1986. Base neutralizing capacity of sediments from an acidic lake. *Water Air Soil Pollut.* 27:297-304.

Molot, L.A., J.G. Hamilton, and G.M. Booth. 1984. Ontario's experimental lake neutralization project: Calcite additions and short-term changes in lake chemistry, in: Lake and Reservoir Management. EPA 440/5-84-001, U.S. Environmental Protection Agency, Washington, D.C., pp. 356-359.

Molot, L.A., J.G. Hamilton, and G.M. Booth. 1986. Neutralization of acidic lakes: Short term dissolution of dry and slurried calcite. *Water Res.* 20:757-761.

Molot, L.A., P.J. Dillon, and G.M. Booth. 1990a. Whole-lake and nearshore water chemistry in Bowland Lake, before and after treatment with $CaCO_3$. *Can. J. Fish. Aquat. Sci.* 47:412-421.

Molot, L.A., L. Heintsch, and K.H. Nicholls. 1990b. Response of phytoplankton in acidic lakes in Ontario to whole-lake neutralization. *Can. J. Fish. Aquat. Sci.* 47:422-431.

Muniz, I.P., and Leivestad. 1980. Acidification—effects on freshwater fish, in: D. Drablos and A. Tollan, eds. Ecological Impact of Acid Precipitation: Proceedings of an International Conference, March 11–14, 1980, Sandefjord, Norway, pp. 84-86.

Nyberg, P. 1984. Effects of liming on fisheries. *Phil. Trans. R. Soc. Lond.* 305:549-560.

Nyberg, P. 1988. Reclamation of acidified Arctic char [*Salvelinus alpinus* (L.)] lakes in Sweden by means of liming. *Verh. Internat. Verein. Limnol.* 23:1737-1742.

Nyberg, P. 1989. The status of liming activities in Sweden. *Living Lakes News* 4(1):4-7.

Nyberg, P., and E. Thornelof. 1988. Operational liming of surface waters in Sweden. *Water Air Soil Pollut.* 41:3-16.

Nyberg, P., E. Degerman, C. Ekstrom, and E. Hornstrom. 1986. Forsurningskansliga rodingsjoar i syd-och mellansverige (Acid-sensitive Arctic char lakes in southern and central Sweden). Nr. 6-1986, Fiskeristyrelsens Sotvattenslaboratorium (Institute of Freshwater Research of the Swedish National Board of Fisheries), Drottningholm, Sweden (In Swedish with English summary).

Office of Technology Assessment. 1984. Acid rain and transported air pollutants: Implications for public policy. OTA-0-204, Office of Technology Assessment, U.S. Congress, Washington, D.C.

Olem, H. 1982. Coal and coal mine drainage. *J. Wat. Pollut. Control Fed.* 54:990-994.

Olem, H., and J.J. Longaker. 1981. Treatment of acid water at abandoned strip mines, in: Proceedings of the ASCE Environmental Engineering Division Specialty Conference, Atlanta, GA, July, 1981, pp. 391-399.

Olem, H., R.W. Brocksen, and T.B. Adams. 1990. Liming lakes takes new twists. *Lake Line* 10(2):2-3, 20-21.

Olsson, H. 1983. Origin and production of phosphatases in the acid Lake Gardsjon. *Hydrobiologia* 93:320.

Omya, Inc. 1989. Product information. Proctor, Vt.

Ostensson, P. 1983. Limestone dosaging equipment at Ryaberg, in: J. Oxley, ed. Proceedings of the Conference: Liming of Running Water, 13 September 1983, Allbogatan, Alvesta, Sweden, Battelle Memorial Institute, Columbus, Oh.

Pearson, F.H., and A.J. McDonnell. 1975a. Use of crushed limestone to neutralize acid wastes. *J. Environ. Eng. Div. A.S.C.E.* 100:139-158.

Pearson, F. H., and A.J. McDonnell. 1975b. Limestone barriers to neutralize acidic streams. *J. Environ. Eng. Div. A.S.C.E.* 101:425-440.

Pearson, F.H., and A.J. McDonnell. 1978. Limestone packed tumbling drums for acidity reduction. *J. Wat. Pollut. Control Fed.* 50:723-733.

Penn Environmental Consultants. 1983. Design manual: Neutralization of acid mine drainage. EPA-600/2-83-001, U.S. Environmental Protection Agency, Cincinnati, Oh., 58 pp.

Pfizer, Inc. 1989. Product information, New York, N.Y.

Porcella, D.B. 1987. Overview of approach and liming of lakes for the Lake Acidification Mitigation Project (LAMP). *Lake Reserv. Manage.* 3:401-403.

Porcella, D.B. 1988. An update on the Lake Acidification Mitigation Project. *Water Air Soil Pollut.* 41:43-52.

Porcella, D.B. 1989. Mitigation of acidic conditions in lakes: An overview of an ecosystem perturbation experiment. *Can. J. Fish. Aquat. Sci.* 46:246-248.

Porcella, D.B., C.L. Schofield, J.V. DePinto, C.T. Driscoll, P.A. Bukaveckas, S.P. Gloss, and T.C. Young. 1989. Mitigation of acidic conditions in lakes and streams. *Adv. Environ. Sci.* 4:159-186.

Powell, M.J. 1977. An assessment of brook trout planting in a neutralized lake as compared to four other Sudbury area lakes. Research Report, Ontario Ministry of Natural Resources, Sudbury, Ontario.

Raddum, G.G., G. Hagenlund, and G.A. Halvorson. 1984. Effects of lime treatment on the benthos of Lake Sondre Boksjo. Report 61, Institute of Freshwater Research, National Swedish Board of Fisheries, Lund, Sweden, pp. 167-176.

Raifaill, B.L., and W.G. Vogel. 1978. A Guide for Vegetating Surface-Mined Lands for Wildlife in Eastern Kentucky and West Virginia. Report FWS/OBS-78/84, U.S. Fish and Wildlife Service, Harpers Ferry, W.V., 89 pp.

Riely, P.L., and D.B. Rockland. 1988. Evaluation of liming operations through benefit-cost analysis. *Water Air Soil Pollut*. 41:293-329.

Ripl, W. 1980. Lake restoration methods developed and used in Sweden, in: Restoration of Lake and Inland Waters: Proceedings of an International Symposium on Inland Waters and Lake Protection, EPA 440/5-81-010, U.S. Environmental Protection Agency, Washington, D.C., pp. 495-500.

Robbins, M.A. 1990. Effects of liming and fish introduction on Chaoborus populations in acidic Adirondack lakes. M.S. Thesis, Indiana University, Bloomington.

Roberts, D.A., and C.W. Boylen. 1989. Effects of liming on the epipelic algal community of Woods Lake, New York. *Can. J. Fish. Aquatic Sci.* 46:287-294.

Rodhe, W. 1981. Reviving acidified lakes. *Ambio*. 10:195-196.

Rosseland, B.O. 1990. Personal communication. Description of Norwegian operational liming program—1983-1989. Norwegian Institute for Water Research, Oslo, Norway.

Rosseland, B.O., and A. Hindar. 1988. Liming of lakes, rivers, and catchments in Norway. *Water Air Soil Pollut*. 41:165-188.

Rosseland, B.O., and O.K. Skogheim. 1984. Attempts to reduce effects of acidification on fishes in Norway by different mitigation techniques. *Fisheries* (Bethesda) 9(1):10-16.

Rosseland, B.O., O.K. Skogheim, H. Abrahamsen, and D. Matzow. 1986. Limestone slurry reduces physiological stress and increases survival of Atlantic salmon *(Salmo salar)* in an acidic Norwegian river. *Can. J. Fish. Aquat. Sci.* 43:1888-1893.

Sandoy, S. 1990. Personal communication. Estimated number of lakes in Norway that may benefit from liming. Directorate for Nature Management, Trondheim, Norway.

Saunders, W.P., D.L. Britt, J.D. Kinsman, J. DePinto, P. Rodgers, S. Effler, H. Sverdrup, and P. Warvinge. 1985. APMP Guidance Manual Volume I: APMP Research Requirements. Biol. Rep. 80(40.24), U.S. Fish and Wildlife Service, Eastern Energy and Land Use Team, 207 pp.

Schaffner, W.R. 1989. The effects of neutralization and addition of brook trout (*Salvelinus fontinalis*) on the limnetic zooplankton communities of two acidic lakes. *Can. J. Fish. Aquat. Sci.* 46:295-305.

Scheffe, R.D., J.V. DePinto, and W.G. Booty. 1986a. Development of methodology for predicting reacidification of $CaCO_3$-treated lakes. *Water Air Soil Pollut.* 31:857-864.

Scheffe, R.D., J.V. DePinto, and K.R. Bilz. 1986b. Laboratory and field testing of dose calculation methods for neutralization of Adirondack lakes. *Water Air Soil Pollut.* 31:799-807.

Scheider, W., and T.G. Brydges. 1984. Whole-lake neutralization experiments in Ontario: A review. *Fisheries* (Bethesda). 9(1):17-18.

Scheider, W., and P.J. Dillon. 1976. Neutralization and fertilization of acidified lakes near Sudbury, Ontario. *Wat. Pollut. Res. Can.* 11:93-100.

Scheider, W., B. Cave, and J. Jones. 1975. Reclamation of acidified lakes near Sudbury, Ontario, by neutralization and fertilization. Ontario Ministry of the Environment Report, Rexdale, Ontario, Canada, 48 pp.

Schofield, C.L., and J. Trojnar. 1980. Aluminum toxicity to fish in acidified waters, in: T. Toribara, M. Miller, and P. Morrow, eds. *Polluted Rain*, Plenum Press, New York, N.Y., pp. 341-366.

Schofield, C.L., S.P. Gloss, and D. Josephson. 1986. Extensive evaluation of lake liming, restocking strategies, and fish population response in acidic lakes following neutralization by liming. EELUT, Interim Progress Report, NEC-86/18, U.S. Fish and Wildlife Service, 117 pp.

Schofield, C.L., S.P. Gloss, B.A. Plonski, and R. Spateholts. 1989. Production and growth efficiency of brook trout (*Salvelinus fontinalis*) in relation to food supply and water quality changes in two Adirondack Mountain lakes (New York) following liming. *Can. J. Fish. Aquat. Sci.* 46:333-341.

Schofield, C.L., D.A. Webster, C.A. Guthrie, and W.A. Flick. 1981. Management of acidified waters. Final Report No. C-164480, New York State Department of Environmental Conservation, Cornell University, Ithaca, N.Y., 82 pp.

Schreiber, R.K. 1988. Cooperative federal-state liming research on surface waters impacted by acidic deposition. *Water Air Soil Pollut.* 41:53-74.

Schreiber, R.K. 1990. Personal communication. Description of results of liming research in Massachusetts, Minnesota, Tennessee, and West Virginia as part of APMP—1986–1989. U.S. Fish and Wildlife Service, Kearneysville, W.V.

Schrieber, R.K., and P.J. Rago. 1984. The federal plan for mitigation of acid precipitation effects in the United States: Opportunities for basic and applied research. *Fisheries* (Bethesda) 9(1):31-35.

Schreiber, R.K., C.D. Goodyear, and W.A. Hartman. 1988. Challenges and progress in the federal/state liming research program. *Lake Line* 8:8.

Shortelle, A.B., and E.A. Colburn. 1987. Ecological effects of liming in a Cape Cod kettle pond: A note for fisheries managers. *Lake Reserv. Manage.* 3:436-443.

Siegfried, C.A., J.W. Sutherland, and S.O. Quinn. 1987. Plankton community response to the chemical neutralization of three acidified waters in the Adirondack Mountain region of New York State. *Lake Reserv. Manage.* 3:444-451.

Simonin, H.A. 1988. Neutralization of acidic waters. *Environ. Sci. Tech.* 22:1143-1145.

Simonin, H.A., S. Kohler, K. Czajkowski, F.R. Beeler, W. Gordon, P. Hulbert, W. Kretzer, P. Neth, M. Pfeiffer, T. Wolfe, J. Nasca, S. Quinn, and J. Sutherland. 1988. Draft generic environmental impact statement on the New York State Department of Environmental Conservation program of liming selected acidified waters. New York State Department of Environmental Conservation, Albany, N.Y. 152 pp.

Smith, R.K. 1985. The influence of sodium carbonate neutralized acid soft water on the survival of *Daphnia magna*. *Lake Reserv. Manage.* 1:119-121.

Staubitz, W.W., and P.J. Zarriello. 1989. Basin and hydrologic characteristics of two headwater Adirondack lakes. *Can. J. Fish. Aquat. Sci.* 46:268-276.

Stoclet, D., ed. 1979. Contracid™ : New lake management and restoration method. Press information, Atlas Copco, Inc., Wayne, N.J., 5 pp.

Stokes, P. 1986. Ecological effects of acidification on primary producers in aquatic ecosystems. *Water Air Soil Pollut.* 30:421-424.

Stross, R.G., and A.D. Hasler. 1960. Some lime-induced changes in lake metabolism. *Limnol. Oceanogr.* 5:265-272.

Stumm, W., and J.J. Morgan. 1981. *Aquatic Chemistry.* John Wiley, New York, N.Y.

Svensson, J., and M. Bratt. 1989. Forsaljning av kalk for jord- och tradgardsbruk samt for kalkning av sjoar och vattendrag aren 1986–1988 (Sales of lime for agricultural and horticultural purposes and for lakes and rivers in 1986–1988). Na 32 SM 8901, Statistiska centralbyran (SCB—Statistics Sweden), Orebro, Sweden (in Swedish with English Summary).

Sverdrup, H.U. 1982a. Lake liming. *Chem. Scr.* 22(1):12-18.

Sverdrup, H.U. 1982b. Lake liming: The calcite utilization efficiency and long-term effect on alkalinity in several Swedish lake liming projects. *Vatten* 34:341-344.

Sverdrup, H.U. 1984. Calcite dissolution and acidification mitigation strategies, in: Lake and Reservoir Management. EPA 440/5-84-001, U.S. Environmental Protection Agency, Washington, D.C., pp. 345-355.

Sverdrup, H.U. 1985. Calcite Dissolution Kinetics and Lake Neutralization. Ph.D. Thesis, Lund Institute of Technology, Lund, Sweden, 169 pp.

Sverdrup, H.U. 1990. Personal communication. Description of Scandinavian liming experiences. Lund Institute of Technology, Lund, Sweden.

Sverdrup, H.U., and I. Bjerle. 1982. Dissolution of calcite and other related minerals in acidic aqueous solution in a pH-stat. *Vatten* 38:59-73.

Sverdrup, H.U., and I. Bjerle. 1983. The calcite utilization efficiency and the long term effect of alkalinity in several Swedish lake liming projects. *Vatten* 39:41-54.

Sverdrup, H., and P. Warfvinge. 1985. A reacidification model for acidified lakes neutralized with calcite. *Water Resour. Res.* 21:1374-1380.

Sverdrup, H., and P. Warfvinge. 1988a. Lake liming in different types of acid lakes using various types of calcite powders and methods. *Water Air Soil Pollut.* 41:189-222.

Sverdrup, H.U., and P.G. Warfvinge. 1988b. What is left for researchers in liming? A critical review of state-of-the-art acidification mitigation. *Lake Reserv. Manage.* 4:87-97.

Sverdrup, H., R. Rassmussen, and I. Bjerle. 1984. Simple model for the reacidification of limed lakes, taking the simultaneous deactivation and dissolution of calcite in the sediments into account. *Chem. Scr.* 24:53-66.

Sverdrup, H.U., P.G. Warfvinge, and J. Fraser. 1985. The dissolution efficiency for different stream liming methods and technologies. *Vatten.* 41:155-163.

Sverdrup, H.U., P.G. Warfvinge, and I. Bjerle. 1986. A simple method to predict the time required to reacidify a limed lake. *Vatten.* 42:10-15.

Swedish Environmental Protection Board. 1988. Kalkning av sjoar och vattendrag (Liming of lakes and streams). Naturvardsverket (Swedish Environmental Protection Board), Solna, Sweden, 74 pp. (In Swedish).

Swedish Fishing Journal. 1909. Kalkgidning i fiskdammar (Liming of fish ponds). *Svensk Fiskeritidskrift* 18:120-122 (in Swedish).

Swedish Ministry of Agriculture Environment Committee. 1982. Acidification today and tomorrow. Risbergs Tryckeri AB, Uddevalla, Sweden. 231 pp.

Tervet, D.J., and R. Harriman. 1988. Changes in pH and calcium after selective liming in the catchment of Loch Dee, a sensitive and rapid-turnover loch in southwest Scotland. *Aquaculture Fish. Manage.* 19:191-203.

Theiss, R. 1981. An Evaluation of the Recovery of an Adirondack

Acid Lake Through Chemical Manipulation. M.S. Thesis, Clarkson College, Potsdam, N.Y. 112 pp.

Thornelof, E. 1990. Personal communication. Description of Swedish operational liming program—1982–1989. Swedish Environmental Protection Board, Solna, Sweden.

Thornton, K.W., D. Marmorek, and P. Ryan. 1990. Methods for projecting future changes in surface water acid-base chemistry. State of Science and Technology Report 14, in: National Acid Precipitation Assessment Program, Acidic Deposition: State of Science and Technology, Volume II, September 1990.

Thunberg, B. 1988. Liming is just a 'holding operation.' *Acid Mag.* 7:8-9.

Turner, R.S., R.B. Cook, H. V. Miegroet, D.W. Johnson, J.W. Elwood, O.P. Bricker, S.E. Lindberg, and G.M. Hornberger. 1990. Watershed and lake processes affecting chronic surface water acid-base chemistry. State of Science and Technology Report 10. In: National Acid Precipitation Assessment Program, Acidic Deposition: State of Science and Technology, Volume II, September 1990.

U.S. Fish and Wildlife Service. 1982. The liming of acidified waters: Issues and research needs. FWS/OBS-80/40.14, U.S. Fish and Wildlife Service, Eastern Energy and Land Use Team, Washington D.C.

Underwood, J., A.P. Donald, and J.H. Stoner. 1987. Investigations into the use of limestone to combat acidification in two lakes in west Wales. *J. Environ. Manage.* 24:29-40.

Warfvinge, P.G., and H.U. Sverdrup. 1988a. Watershed liming. *Lake Reserv. Manage.* 4:899-106.

Warfvinge, P., and H. Sverdrup. 1988b. Soil liming as a measure to mitigate acid runoff. *Water Resour. Res.* 24:701-712.

Warfvinge, P., and H. Sverdrup. 1989. Modeling limestone dissolution in soils. *Soil Sci. Soc. Am. J.* 53:44-51.

Waters, T.F. 1956. The effects of lime application to acid bog lakes in northern Michigan. *Trans. Am. Fish. Soc.* 86:329-344.

Waters, T.F., and R.C. Ball. 1957. Lime application to a soft-water, unproductive lake in northern Michigan. *J. Wildl. Manage.* 21:385-391.

Watt, W.D., G.J. Farmer, and W.J. White. 1984. Studies on the use of limestone to restore Atlantic salmon habitat in acidified rivers, in: Lake and Reservoir Management. EPA 440/5-84-001, U.S. Environmental Protection Agency, Washington, D.C., pp. 374-379.

Watt, W.D. 1986. The case for liming some Nova Scotia salmon rivers. *Water Air Soil Pollut.* 31:775.

Weatherley, N.S. 1988. Liming to mitigate acidification in freshwater ecosystems: A review of the biological consequences. *Water Air Soil Pollut.* 39:421-437.

Weatherley, N.S., S.P. Thomas, and S.J. Ormerod. 1989. Chemical and biological effects of acid, aluminum and lime to a Welsh hill-stream. *Environ. Pollut.* (G.B.) 56:283-297.

Welsh Water. 1986. The survival and physiological responses of brown trout stocked in two acidic upland lakes before and after liming and an assessment of the performance of trout fisheries re-established in them. Report SW/86/30, Welsh Water, Dyfed, Wales, U.K.

White, W.J., W.D. Watt, and C.D. Scott. 1984. An experiment on the feasibility of rehabilitating acidified Atlantic salmon habitat in Nova Scotia by the addition of lime. *Fisheries* (Bethesda) 9(1):25-30.

Wilmoth, R.C. 1977. Limestone and lime neutralization of ferrous iron acid mine drainage. EPA-600/2-77-101, U.S. Environmental Protection Agency, Cincinnati, Oh.

Wright, R.F. 1984. Changes in the chemistry of Lake Hovvatn, Norway, following liming and reacidification. Report No. 0-

80044-01, Norwegian Institute for Water Research, Oslo, Norway, 68 pp.

Wright, R.F. 1985. Chemistry of Lake Hovvatn, Norway, following liming and reacidification. *Can. J. Fish. Aquat. Sci.* 42:1103-1113.

Yan, N.D. 1983. Effects of changes in pH on transparency and thermal regimes of Lohi Lake, near Sudbury, Ontario. *Can. J. Fish. Aquat. Sci.* 40:621-626.

Yan, N.D., and P.J. Dillon. 1984. Experimental neutralization of lakes near Sudbury, Ontario, in: J. Nriagu, ed. *Environmental Impacts of Smelters*. John Wiley & Sons, Toronto, Canada.

Yan, N.D., and P. Stokes. 1978. Phytoplankton of an acidic lake, and its responses to experimental alterations of pH. *Environ. Conserv.* 5:93-100.

Yan, N., W. Scheider, and P. Dillon. 1977a. Chemical and biological changes in Nelson Lake, Ontario, following experimental elevation of lake pH. *Wat. Pollut. Res. Can.* 12:213-231.

Yan, N.D., J. Jones, B. Cave, L. Scott, and M. Powell. 1977b. The effects of experimental elevation of lake pH on the chemistry and biology of Nelson Lake near Sudbury, Ontario. Ministry of the Environment, Sudbury, Ontario.

Yan, N.D., R.E. Girard, and C.J. Lafrance. 1979. Survival of rainbow trout, *Salmo gairdneri*, in submerged enclosures in lakes treated with neutralizing agents near Sudbury, Ontario. Technical Report LTS 79-2, Ministry of the Environment, Sudbury, Ontario.

Yocum, S. 1976. Trough Creek limestone barrier installation and evaluation. EPA-600/2-76-114, U.S. Environmental Protection Agency, Cincinnati, Oh, 93 pp.

Young, T.C., J.V. DePinto, J.R. Rhea, and R.D. Scheffe. 1989. Calcite dose determination, treatment efficiency and sediment dose response to whole lake neutralization. *Can J. Fish. Aquat. Sci.* 46:315-322

Young, T.C., J.R. Rhea, and G. McLaughlin. 1986. Characterization and neutralization requirements of two acidic lake sediments. *Water Air Soil Pollut.* 31:839-846.

Zengerle, M.W., and M.A. Allan. 1987. Status reports on selected environmental issues. Volume 1: Liming of acidified waters. EA-5097-SR, Electric Power Research Institute, Palo Alto, Ca., 12 pp.

Zurbuch, P. 1963. Dissolving limestone from revolving drums in flowing waters. *Trans Am. Fish. Soc.* 92:173-178.

Zurbuch, P. 1984. Neutralization of acidified streams in West Virginia. *Fisheries* (Bethesda) 9(1):42-47.

Appendix A

Surface Waters Mentioned in Book

Table A.1 Location and Characteristics of Lakes Mentioned Throughout Book

Lake	Location	Surface Area (ha)	Year Treated	Quantity (Tonnes)	Material	Application Method	References	Page(s)
Asnacomet Pond	Hubbardston, Ma.	420.0	1982	45.4	Agricultural limestone	Boat	Bergin, 1984 Fraser et al., 1985	56
Berry Pond	North Andover, Ma.	1.2	1957	1.1	Agricultural limestone		Bergin, 1984	10
Black Forest	West Germany					Watershed	Brocksen et al., 1988	21
Bowland Lake	Sudbury, Ontario Canada	108	1983	83	Ultra-fine limestone	Fixed-wing aircraft	Booth et al., 1987; Molot et al., 1984	21 68 149
Cather Lake	Wisconsin	3.4	1950		Agricultural limestone	Boat	Fraser and Britt, 1982	11
Cranberry Pond	Adirondacks, N.Y.		1985		Limestone slurry	Helicopter	Young et al., 1989	148
Gabriel Lake	Wisconsin		1969		Hydrated lime		Garrison, 1984	12

Lake Gardsjon	Sweden	31.2					Dickson, 1985, 1988	23 154
Lake Hannah	Ontario, Canada				Limestone	Boat	Scheider et al., 1975	168
Lake Hovvatn	Norway	114	1980	200	Limestone	Manure spreader over ice	Fraser and Britt, 1982, Wright, 1985	23 149
Lake Lilla Galtsjon	Sweden	14.7	1980	40	Sodium carbonate	Sediment injection	Lindmark, 1982, 1985	63 151
Lake Lohi	Ontario, Canada				Limestone	Boat	Yan, 1983	148
Lake Masen	Halland, Sweden	420	1984	1700	Limestone	Barge/ pressure tank	Fraser et al., 1985	59
Lake Ommern	Southwest Sweden				Limestone		Hornstrom and Ekstrom,1986	169
Lake Stensjon	Sweden						Henricksen et al., 1984	171
Lake Stora Harsjon	Sweden						Hornstrom and Ekstrom, 1986	171

Table A.1 Location and Characteristics of Lakes Mentioned Throughout Book (continued)

Lake	Location	Surface Area (ha)	Year Treated	Quantity (Tonnes)	Material	Application Method	References	Page(s)
Lake Stora Holmevatten	Sweden						Hasselrot et al., 1987	153
Lake Stora Lee	Varmland, Sweden	8500	1982	9000	Limestone	Barge/ pressure tank	Sverdrup, 1984	61
Lake Tvarsjon	Sweden						Hornstrom and Ekstrom,1986	171
Little Hell Lake	Fort Devens, Mass.	2.4	1963	3.2	Agricultural limestone		Bergin, 1984 Fraser and Britt, 1982	10
Little Simon Pond	Adirondacks, N.Y.		1985		Limestone	Helicopter	Porcella, 1989	140
Loch Dee	Scotland		1984		Agricultural limestone	Watershed	Tervet and Harriman, 1988	24
Loch Fleet	Scotland	17.0	1986	350	Limestone powder, slurry, and pellets	Watershed	Brown et al., 1988	24 98 149

Middle Lake	Ontario, Canada				Limestone	Boat	Scheider et al., 1975	168
Nelson Lake	Ontario, Canada					Boat	Yan and Dillon, 1984	169, 180
Peter-Paul Lake	Michigan		1951		Calcium hydroxide (Peter Lake)	Boat	Stross and Hassler, 1960	12
Pollen Pond	Norway	4.6	1980	40	Limestone	Manure spreader over ice	Wright, 1985	23, 149
Sandy Lake	Bedford, Nova Scotia, Canada	74	1981	135	Limestone	Slurry from boat	Watt et al., 1984 Watt, 1986 White et al., 1984	22 58 181
Sherwood Lake	West Virginia		1964–69		Limestone	Upstream rotary drum	Zurbuch, 1984	92
Spruce Knob Lake	West Virginia	11.3	1958		Agricultural limestone		Zurbuch, 1963, 1984 Fares and Kinsman, 1984	11
Summit Lake	West Virginia	17.4	1965		Agricultural limestone		Zurbuch, 1963, 1984 Fares and Kinsman, 1984, Fraser and Britt, 1982	11

Table A.1 Location and Characteristics of Lakes Mentioned Throughout Book (continued)

Lake	Location	Surface Area (ha)	Year Treated	Quantity (Tonnes)	Material	Application Method	References	Page(s)
Thrush Lake	Minnesota		1988	4.5	Limestone	Slurry from boat	Schreiber, 1988	19 147
Tjonnstrond Lake	Norway	25.0	1983	75.0	Limestone helicopter	Watershed	Rosseland and Hindar, 1988	100 149
Trout Lake	Sudbury, Ontario, Canada	290	1984	170	Limestone	Fixed-wing aircraft	Fraser et al., 1985	21 52
Turk Lake	Wisconsin	3.5	1950		Agricultural limestone	Boat	Garrison, 1984	11
Wallum Lake	Rhode Island/ Ma. Border	130	1987	99.5	Limestone	Barge	Brocksen and Emler, 1988; Adams and Brocksen, 1988	61
White Deer Lake	Pike County, Pa.	19.8	1987	16.1	Limestone	Barge/ pressure tank	Brocksen and Emler, 1988; Adams and Brocksen, 1988	60 162

Woods Lake	Adirondacks, N.Y.	25.0	1985	22.7	Slurried	Helicopter limestone	Scheffe et al., 1986b	21 67 148
Woods Lake	Adirondacks, N.Y.	25.0	1989	1110	Pelletized limestone to watershed	Helicopter	Boutacoff, 1990 Olem et al., 1990	100
Wolf Pond	Franklin County, N.Y.	20	1987		Sodium bicarbonate and hydrated lime		Siegfried et al., 1987	41

Table A.2. Location and Characteristics of Flowing Waters Mentioned Throughout Book

Stream/River	Location	Average Flow (m^3/sec)	Year Treatment Begun	Average Dosage (kg/h)	Material	Application Method	References	Page(s)
Bacon Ridge Branch	Maryland		1986		Limestone slurry	Automated doser	Greening et al., 1987; Janicki and Greening, 1988a,b	20
Condon Run	Elkins, WV				Limestone		Zurbuch, 1984	11
Constable Creek/ Big Moose Lake	Adirondacks, NY				Limestone	Limestone barrier	Driscoll et al., 1982	95
Dogway Fork	West Virgina		1988		Limestone aggregate	Rotary drum	Zurbuch, 1984	19 202
Fyllaen River	Ryaberg, Sweden	1.5–15	1982		Limestone	Electric dry powder doser	Fraser et al., 1985	79
Gifford Run	Pennsylvania		1976		Limestone	Limestone barrier	Arnold, 1980, 1981	95
Granny Finley Branch of Chester River	Monongahela Nat. Forest, WV		1986		Limestone	Wire mesh barrier	Greening et al., 1987; Janicki and Greening, 1988a,b	20

Laurel Branch	Tennessee		1989		Limestone	Water-powered doser	Schreiber, 1988	202
Lower Great Brook	Nova Scotia, Canada		1980		Limestone aggregate	Limestone barrier	Fraser et al., 1982	97
Mattawoman Creek	Maryland		1987		Limestone slurry	Automated doser	Greening et al., 1987; Janicki and Greening, 1988a,b	20 79
Otter Creek	Monongahela Natl.Forest, W.V.	0.113	1964	5.4	Limestone aggregate	Rotary drum	Zurbuch, 1984	20
Otter Creek	Monongahela Natl. Forest, W.V.		1980		Limestone aggregate	Self-feeding system	Zurbuch, 1984	20 91
Rock Branch of Patuxent River	Maryland		1986		Limestone	Wire mesh barrier	Greening et al., 1987 Janicki and Greening, 1988a,b	20
Silkbacken	Hallefors, Sweden	0.54	1985	4.0-9.9	Limestone	Water-powered doser	Fraser et al., 1985	86

Table A.2. Location and Characteristics of Flowing Waters Mentioned Throughout Book (continued)

Stream/River	Location	Average Flow (m^3/sec)	Year Treatment Begun	Average Dosage (kg/h)	Material	Application Method	References	Page(s)
Stocketts Run of Patuxent River	Maryland		1988		Limestone	Propane-powered doser	Greening et al., in press	20
Whetstone Brook	Massachusetts		1989		Limestone	Water-powered doser	Schreiber, 1988	202
Yocum Run	West Virginia		1986		Limestone powder	Stream bottom	Brocksen and Emler, 1988	210

Appendix B

Comparison of Limestone Dose Calculations and Prediction of Reacidification Rates

The dose calculation and reacidification predictions from the DeAcid model (Fraser et al., 1985) are compared to the empirical relations given in the Swedish Liming Handbook (Swedish Environmental Protection Board, 1988) in this appendix.

CHARACTERISTICS OF EXAMPLE LAKES

Identical parameters are used for each model from four hypothetical lakes. The characteristics of these systems are based on typical conditions in small and large surface waters in the Adirondack region of New York. Table B.1 summarizes the characteristics that are required for computing the estimated dose for the four lakes. Michael's Lake has a pH of 5.3, an ANC of -10 μeq/L, and a lake retention time of 2.0 years. Lake Louis has the same water chemistry, but has a shorter retention time (0.75 year). Lake Sheila Anne has a retention time of 2.0 years, but is much more acidic (pH = 4.7; an ANC = -50 μeq/L). Bella Pond is similar to Michael's Lake in chemistry and lake retention time, but is much smaller (10 ha).

RESULTS AND DISCUSSION

Table B.2 summarizes the outputs from each technique for the example lakes. The DeAcid and Swedish Liming Handbook

Table B.1 Characteristics of Example Lakes for Comparing Limestone Dose Requirements Calculated by Two Techniques[a]

Lake Name	Surface Area (ha)	Volume (10^6 m^3)	Water Retention Time (years)	ANC (µeq/L)	pH
Michael's Lake	30	1.0	2.0	-10	5.3
Lake Louis	30	1.0	0.75	-10	5.3
Lake Sheila Anne	30	1.0	2.0	-50	4.7
Bella Pond	10	0.3	2.0	-10	5.3

[a] For all lakes, calcium and DOC were both 1.0 mg/L. The application method was slurry applied from a boat and the base material was 0–0.2 mm high-calcium limestone. See Table B.2 for the calculation techniques used.

Table B.2 Results of Limestone Dose Estimates for Example Lakes (Table B.1) Calculated by Two Techniques, the DeAcid model and the Swedish Liming Handbook

	Limestone Dose, Tonnes		Reacidification Time, Years	
Lake	DeAcid Model	Swedish Liming Handbook	DeAcid Model	Swedis Liming Handbook
Michael's Lake	25	20	5.0	3.5–5.0
Lake Louis	25	30	2.0	2.0–3.0
Lake Sheila Anne	22	30	3.0	3.5–5.0
Bella Pond	8	6	2.1	3.5–5.0

methods provide very similar dose estimates and reacidification predictions for Michael's Lake and Lake Louis. Lake Louis, with a shorter retention time, requires the same dose as Michael's Lake when DeAcid is applied and a larger dose when the Swedish method is used. The Swedish method apparently calls for larger doses for lakes with shorter retention times so that the time to reacidification is extended.

The two methods do not agree when they are applied to the more acidic Lake Sheila Anne. The DeAcid method calls for 22 tonnes of limestone and the Swedish method for 30 tonnes. The dose, as calculated by this model, is lower when DeAcid is applied to the pH 4.7 lake than when it is applied to the pH 5.3 lake. This may be partly due to the increased limestone dissolution efficiency of the more acidic lake (36% vs. 23%).

As expected, the DeAcid model predicts a shorter time to reacidification for the more acidic lake. The Swedish method considers only lake retention time to calculate reacidification and does not take lake water chemistry into consideration (see Figure 2.36). This model, therefore, predicts the same range of 3.5–5 years to reacidification for the identical lakes with different chemistries.

Bella Pond, with a smaller lake surface area and volume than the other three lakes, requires a much smaller limestone dose. The DeAcid and Swedish models calculate similar limestone doses (7.8 and 6 tonnes, respectively).

SUMMARY

The two models yield similar estimations of doses for the example lakes, but do not agree for all scenarios. The time for lake reacidification when DeAcid is used varies consistently with differences in lake retention time and lake pH and ANC. The Swedish Liming Handbook model accounts for lake pH and ANC changes in calculating dose, but considers only lake water retention in estimating reacidification.

Appendix C

Results of Major Mitigation Projects in the United States

Only a few of the surface waters that have been limed in the United States were later evaluated in sufficient detail to provide information on the chemical and biological effects of liming. Projects sponsored by two not-for-profit private organizations (Electric Power Research Institute and Living Lakes, Inc.), the U.S. Fish and Wildlife Service, and two state agencies (Maryland and West Virginia) provide a representative range of evaluations to date on the effects of liming in the United States. Many of the study sites were jointly funded by the organizations; these are in the subsection on the organization that had the lead role in planning and data collection.

Most of the programs involved lake liming; only recently have a number of projects been initiated on streams. No information is available on watershed liming. Liming the Woods Lake watershed in New York was conducted in October 1989 and results are not yet available.

ELECTRIC POWER RESEARCH INSTITUTE LAKE ACIDIFICATION MITIGATION PROGRAM

The Electric Power Research Institute has conducted aquatic liming exclusively in the Adirondack region of New York State. The study is ongoing and focuses on research and demonstration of effective lake liming techniques (Porcella, 1988, 1989). This study provided the most comprehensive evaluation to date on the effects of liming lakes in the United States.

Three lakes were limed during 1985–1986. Extensive data collection efforts were undertaken before, during, and immediately after liming to monitor changes in water quality, sediment chemistry, and aquatic biota. Sampling was continued in 1987 and 1988 to provide information on chemical and biological changes during reacidification. Only comprehensive information on the two lakes limed in 1985—Cranberry Pond and Woods Lake—has been reported to date (Driscoll et al., 1989a; Driscoll et al., 1989b; Staubitz and Zarriello, 1989; Gloss et al., 1989a; Roberts and Boylen, 1989; Schaffner, 1989; Fordham and Driscoll, 1989; Young et al., 1989; Schofield et al., 1989; Evans, 1989; Bukaveckas, 1989). The third lake, Little Simon Pond, was limed in 1986, but few results have yet been reported (Bukaveckas, 1988b).

Morphometric characteristics of Cranberry Pond and Woods Lake are shown in Table C.1. The two lakes were characterized as small acidic headwater systems with short residence times (Staubitz and Zarriello, 1989). The lakes differ in size and depth but have similar watershed characteristics (Figure C.1).

The fine limestone slurry distributed to the lakes resulted in high dissolution efficiency (Driscoll et al., 1989b). Four weeks after liming, dissolution was 86% in Woods Lake and 79% in Cranberry Pond. Essentially all of the limestone was dissolved in both lakes within 4 months after liming; little accumulation of limestone in bottom sediment was observed.

Water chemistry changes in Woods Lake before and after liming are shown in Figure C.2. Short-term changes in water chemistry included an immediate increase in pH from less than 5.0 to above 9.0; a stabilization in pH below 8.0 after equilibration with atmospheric CO_2; increases in calcium, ANC, and DIC; and a shift in speciation of aluminum from a dominance of organic (nonlabile) monomeric aluminum to the inorganic (labile) monomeric form (Fordham and Driscoll, 1989; Driscoll et al., 1989a). After about 1 month, aluminum, manganese, and zinc in the water column were markedly reduced (Driscoll et al., 1989a). The metals accumulated as mineral precipitates in lake bottom sediments.

Young et al. (1989) evaluated the fate of calcium in lake sediments. Sediment pH was higher in the surface sediments and decreased progressively to depths where sediment (>6cm) remained unchanged after neutralization. Calcium increased in surficial sediments as sediment base neutralizing capacities de-

Table C.1 Morphometric Characteristics of Cranberry Pond and Woods Lake (Source: Staubitz and Zarriello, 1989)

Variable	Woods Lake	Cranberry Pond
Surface area (ha)	8	23
Watershed area (ha)	155	207
Volume (m^3)	207,000	800,000
Maximum depth (m)	7.6	12
Mean depth (m)	2.9	3.5
Retention time (mo)	2.1	5.8

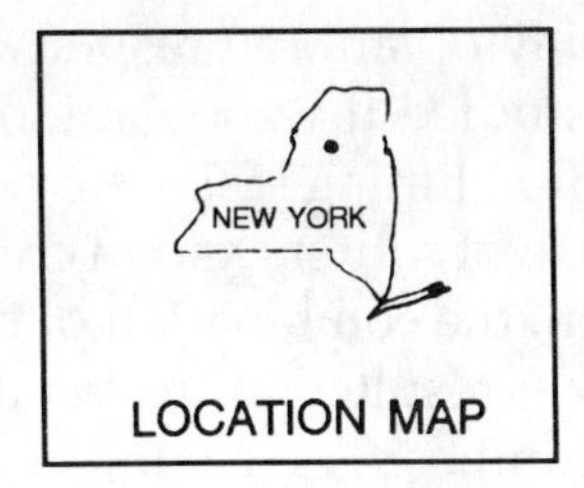

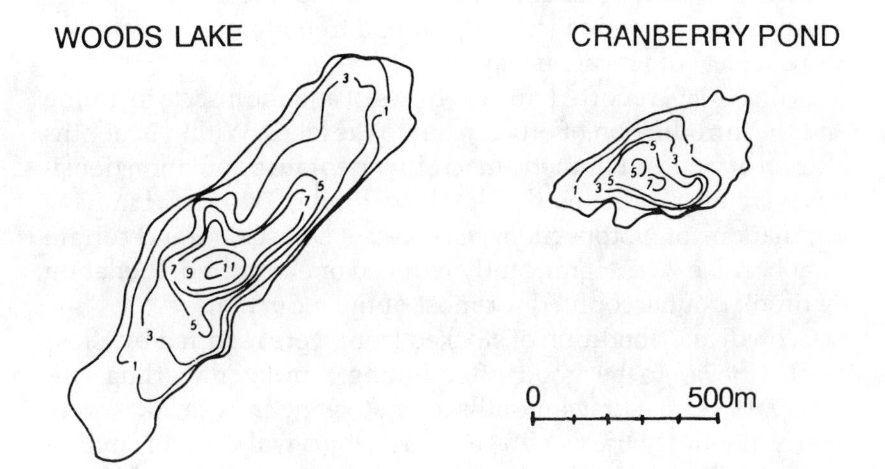

Figure C.1 Woods Lake and Cranberry Pond. Numbers refer to lake depth in meters. (Modified from Staubitz and Zarriello, 1989.)

creased. Over the 16-month period between treatment and reacidification, sediment chemical characteristics returned to pretreatment conditions as alkalinity was released to the overlying water. The increases in ANC from sediments, however, were of minor importance to the postponement of reacidification. The rate of reacidification was largely explained by the flushing of ANC from the lakes by incoming water during the short residence times (Driscoll et al., 1989a; Staubitz and Zarriello, 1989; Gloss et al., 1989a).

Phytoplankton community characteristics were studied before and after liming to the two lakes (Bukaveckas, 1989). Significant reductions in phytoplankton density were observed after liming. These changes appeared to be due to low CO_2 levels immediately after treatment, as indicated by the initially high pH values. Over the longer term, nutrients did not appear to be limiting. Roberts and Boylen (1989) reported sediment algae responses to liming to be consistent with phytoplankton responses.

Schaffner (1989) studied the zooplankton communities in the lakes before and after liming. Short-term changes included a reduction in most taxa of rotifers and in crustacean communities. Over the longer term, the combination of liming and stocking of brook trout in the lakes resulted in an overall positive effect on the zooplankton communities. For example, a shift was observed toward numerically smaller rotifer communities, more species having shared dominance, indicating a trend toward conditions more typical of less acidic lakes.

Evans (1989) reported the responses of aquatic insects to liming and the introduction of brook trout to the lakes. Within 3 months after treatment, all limnetic macroinvertebrate population densities were near or below the detection limit of $0.01/m^3$. Limnetic populations of Notonectidae, Corixidae, Dytiscidae, and certain Chaoboridae were significantly reduced or eliminated. Predation by brook trout accounted for most of the reductions.

Growth and condition of stocked trout were reported by Gloss et al. (1989a) to be good after liming. Spring fingerling fish survival over the first 4 months after stocking in both lakes were nearly identical (66% vs. 64%) to average survival rates for circumneutral Adirondack lakes. Survival of fish stocked in Woods Lake, however, was lower than that in 10 limed Adirondack waters reported by Schofield et al. (1986). Gloss et al. (1989a) considered this to be due to early summer stocking and incomplete base chemical mixing conditions in Woods Lake.

Large invertebrates made up much of the initial diet for stocked fish. Corresponding reductions in invertebrate populations resulted in a shift in diet to zooplankton. This observation suggested

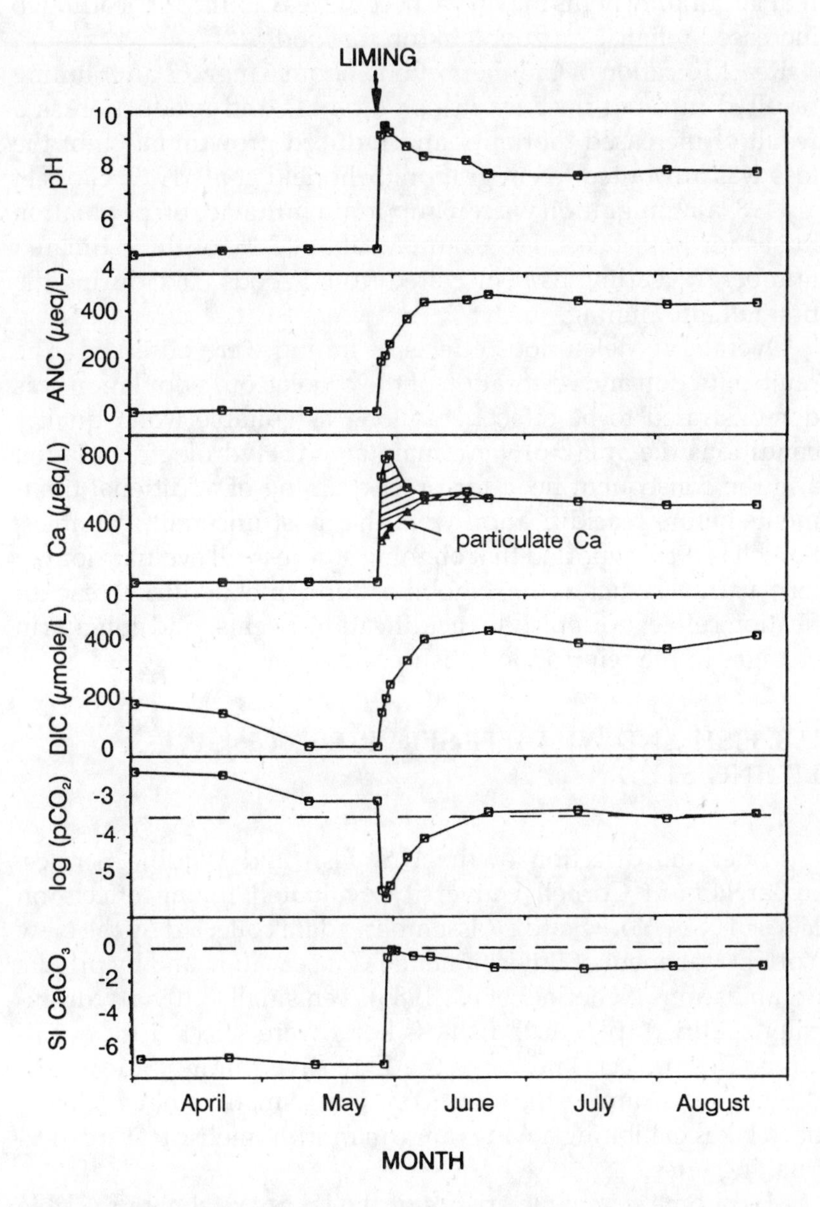

Figure C.2 Epilimnetic volume-weighted changes in water chemistry in Wood Lake before and after liming. (Modified from Fordham and Driscoll, 1989.)

that larger prey items may have become less abundant, leading to increased reliance on zooplankton for food.

Reacidification of Cranberry Pond about 6 months after liming resulted in rapid declines in fish biomass and production as a result of increased mortality and reduced growth; much of the loss was attributed to emigration (Schofield et al., 1989; Gloss et al., 1989a). Emigration was an important influence on population losses for both lakes. For example, about 30% of the estimated number of yearling fish emigrated from Woods Lake during the first fall after liming.

Overall, no deleterious effects of liming were observed. The reintroduction and restoration of the brook trout population was demonstrated to be effective as long as suitable water quality conditions (i.e., pH >6) were maintained. Hydrologic variation and the consequent need for precise timing of additional treatments before reacidification were the most uncertain variables. Porcella (1989) reported that ongoing studies will evaluate longer term variables such as the remobilization of metals during reacidification, effects of episodic acidification events, and long-term changes in the reintroduced fish population.

U.S. FISH AND WILDLIFE SERVICE EXTENSIVE LIMING STUDY

Under the direction of the U.S. Fish and Wildlife Service, researchers at Cornell University evaluated liming effects on lakes in New York State by assembling data collected by the New York Department of Environmental Conservation and by private organizations (Schofield et al., 1986). Ten small (<10 ha), chronically acidified (pH <5.0), fishless lakes were selected for evaluation. The project, known as the Extensive Liming Study, was designed to examine the effects of liming small, remote Adirondack lakes exhibiting a wide range in morphometric features and flushing rates.

Morphometric characteristics of the lakes are shown in Table C.2. After the lakes were limed from fixed-wing aircraft, transparency decreased temporarily and thermal stratification changed in one lake, but not in the others.

Chemical changes after liming included increased pH, calcium, and ANC. A summary of the water chemistry in the study lakes

Table C.2 Morphometric Characteristics of Extensive Liming Study Lakes in New York (Source: Schofield et al., 1986)

Lake	Surface Area (ha)	Watershed Area (ha)	Volume (m^3)	Mean Depth (m)	Retention Time (mo)
Mountain	6.0	49.7	278,743	4.7	9.0
Big Chief	1.2	19.5	53,020	4.4	4.6
Little Rock	4.1	179.5	55,105	1.4	0.5
Highrock	4.0	16.7	140,163	3.5	12.3
Trout	3.7	67.5	46,016	1.3	1.2
Barto	5.5	72.5	156,313	2.9	3.1
Jones	5.3	16.8	336,507	6.4	27.9
Indigo	5.7	21.0	207,750	3.7	14.2
Pocket	1.2	23.6	34,718	2.9	2.5
Silver Dollar	0.5	9.8	20,596	4.1	3.6

before and after liming (Table C.3) indicated that aluminum showed little immediate response to the liming treatments. The aluminum species, however, changed from a dominance of labile (inorganic) forms shortly after liming to nonlabile (organic) forms after several months. This was the opposite to what was observed after the treatment of Woods Lake and Cranberry Pond (Fordham and Driscoll, 1989). The shift in aluminum to a less toxic form, coupled with low water temperatures, was reported to contribute to high survival of caged and stocked trout in the study lakes.

High ANC was noted in hypolimnetic waters after liming. This observation was associated with incomplete spring and fall turnover in the lakes and did not benefit stocked trout because dissolved oxygen was low in the hypolimnion. Reacidification rates in the lakes were consistent with water retention times except in two lakes, where partial mixing of hypolimnetic waters prolonged reacidification. Complete loss of fish populations through mortality and emigration was due to the reacidification of these lakes.

The survival of stocked trout was relatively high (40%–80%) in the lakes that did not reacidify during the first year after liming. The maintenance of standing crop levels of 10–20 kg/ha of stocked trout in these lakes for extended periods were predicted to result in significant decreases in growth as a result of decreased food availability. This prediction was based on observations from the Woods Lake and Cranberry Pond study (Gloss et al., 1989b), in

Table C.3 Mean Chemical Characteristics of Extensive Liming Study Lakes in New York Before and After Liming (Source: Gloss et al., 1989a)

	Before Liming					After Liming[a]				
Lake	pH	ANC, μeq/L	Ca, mg/L	Al_l[b] μg/L	Al_t[c] μg/L	pH	ANC, μeq/L	Ca, mg/L	Al_l[b] μg/L	Al_t[c] μg/L
Mountain	4.7	-5.8	1.4	370	1025	6.5	88	4.2	225	407
Big Chief	4.8	0.7	2.1	279	316	6.8	113	4.4	276	488
Little Rock	4.8	3.6	1.0	122	304	6.1	11	2.2	418	568
Highrock	—	-3.0	1.8	183	185	6.6	132	5.1	230	359
Trout	4.8	-3.0	1.8	—	—	6.8	148	5.3	237	391
Barto	4.7	-10.1	0.8	300	413	6.6	111	3.6	184	368
Jones	5.0	-0.8	1.2	86	158	7.1	118	3.8	78	159
Indigo	4.9	0.5	1.1	27	110	6.9	116	3.6	63	140
Pocket	4.3	-35.3	0.7	420	550	6.1	90	4.0	266	561
Silver Dollar	4.3	-40.9	0.7	356	473	6.4	109	4.0	187	430

[a] Mean water chemistry results for 6- to 8-month period immediately after liming.
[b] Labile (inorganic) monomeric aluminum.
[c] Total aluminum.

which stocked trout first consumed the larger invertebrate organisms that made up the bulk of the food supply. It was not known whether equilibrium growth conditions would continue after long periods (>5 years) of continued stocking and maintenance liming in these oligotrophic lakes.

U.S. FISH AND WILDLIFE SERVICE ACID PRECIPITATION MITIGATION PROGRAM

The U.S. Fish and Wildlife Service began this study in 1985 to evaluate the ecological effects of experimentally liming one lake in Minnesota and one stream each in Massachusetts, West Virginia, and Tennessee (Schreiber, 1988; Schreiber et al., 1988).

Thrush Lake in Minnesota was limed in 1988 for protection against episodic acidic input from snowmelt. Some preliminary chemical and biological results were obtained from R.K. Schreiber (personal communication, 1990). In May 1988, 4.5 tonnes of powdered limestone were added to the lake with a slurry box device connected to a boat. Morphometric and chemical characteristics of the lake are shown in Table C.4. Because the lake was stratified and a fine grade of limestone was applied, immediate chemical changes were restricted to the top 6 m of the lake throughout most of the summer. The initial change in pH in this water increased markedly, from 6.5 to 9.5. Under these high pH conditions, it took about 2 weeks for all of the limestone to dissolve, and pH shock to phytoplankton resulted.

The undissolved limestone reduced transparency during the 2-week period. This was followed by increased transparency (Secchi depth >10 m). In the preliminary report, no reason was given for the increased water clarity, although reduced phytoplankton biomass was suggested. Even though total phosphorus decreased, the chlorophyll *a* concentration did not change.

Fall turnover resulted in distribution of the added alkalinity throughout the water column. Target increases in pH and ANC were 0.5 and 200 μeq/L. These increases were reached after complete mixing in the fall. Other chemical changes observed included increases in calcium, magnesium, and specific conductance.

Dramatic declines in some pelagic zooplankton species, including *Holopedium gibberum*, were considered to be consistent

Table C.4 Morphometric and Chemical Characteristics of Thrush Lake, Minnesota (Source: R.K. Schreiber, personal communication, 1990)

Variable	Immediately Before Liming[a]	Immediately After Liming[b]
Surface area, ha	6.62	—
Maximum depth, m	14	—
Mean depth, m	6.8	—
Retention time, years	4.5	—
pH[c]	6.67	9.42
ANC, μeq/L[c]	67	427
Calcium, mg/L[c]	1.9	10.6
Magnesium, mg/L[c]	0.7	0.8
Total aluminum, μg/L[c]	21	25
Conductance, μS/cm[c]	20.0	40.6
Dissolved inorganic carbon, mg/L[c]	1.1	2.5

[a] Values measured before liming on May 18, 1988.
[b] Values measured after liming on May 26, 1988. Limestone was added on May 24, 1988.
[c] Samples collected from the epilimnion (2 m) at midlake.

with chemical changes. Other species showed different distribution patterns, but the causes of the changes were not clear.

Two 7-day bioassays were conducted with young-of-the-year brook trout, yearling rainbow trout, and yearling fathead minnows. One bioassay was conducted during spring snowmelt just before liming and the other was conducted during liming. No significant differences in mean survival time were observed for any of the three species in either of the two bioassays.

Dogway Fork in West Virginia was outfitted in December 1988 with a self-feeding rotary drum device for continuous limestone addition. Initial results are not yet available.

The projects in Massachusetts and Tennessee collected preliming water quality and fish data since 1986. Limestone treatments began in 1989 and results are not yet available (R.K. Schreiber, personal communication, 1990).

MARYLAND

Two Coastal Plain streams, Bacon Ridge Branch and Mattawoman Creek, were selected by the Maryland Department of Natural Resources as experimental streams to be treated with limestone slurry during acidic episodes (Greening et al., 1987; Janicki and Greening, 1988a,b; Greening et al., in press). Electrically powered wet-slurry dosers were installed at both sites in 1986. Extensive analysis of water quality and aquatic biota were conducted, but unusually dry weather precluded full evaluation of the dosers.

During 1987 and 1988, the slurry dosers were effective under most conditions. Dosing was occasionally interrupted during precipitation events due to power failures and clogging of discharge pipes. The project included evaluations of water quality upstream and downstream from the dosers, instream bioassay experiments with white perch and yellow perch, and monitoring anadromous fish species in the stream.

In an example of the upstream and downstream water chemistry results during a storm event on Bacon Ridge Branch (Figure C.3), the pH was maintained above 6.1 downstream from the doser during elevated flows over a 40-hr period. Upstream pH values decreased to 5.2 during this event. ANC levels were maintained above 100 μeq/L throughout the storm and calcium levels also increased downstream. Total (organic and inorganic) monomeric aluminum concentrations were maintained below 30 μg/L downstream, and upstream values increased above this level for more than 24 hr. With few exceptions, water quality at the downstream stations on Bacon Ridge Branch and Mattawoman Creek was maintained during spring anadromous fish spawning season storms at levels near base flows.

Mortality of yellow perch eggs was evaluated during spring at the upstream and downstream sites. The in situ study results indicated that egg mortality at the two sites did not differ significantly. The researchers suggested that the lethal acidity threshold for yellow perch eggs in these streams was below pH 5.8 and eggs were also relatively insensitive to the 8–42 μg/L of total monomeric aluminum present.

Experiments with yolk sac (prefeeding) yellow perch larvae at Bacon Ridge Branch showed that mean larval mortality during storms was significantly higher at the upstream site (69%) compared to the treated downstream site (34%). The findings were attributed only partly to stream pH, because increased turbidity during storms was also considered to cause mortality. However, no significant differences between upstream and downstream sites at Mattawoman Creek were observed. Fish sampling indi-

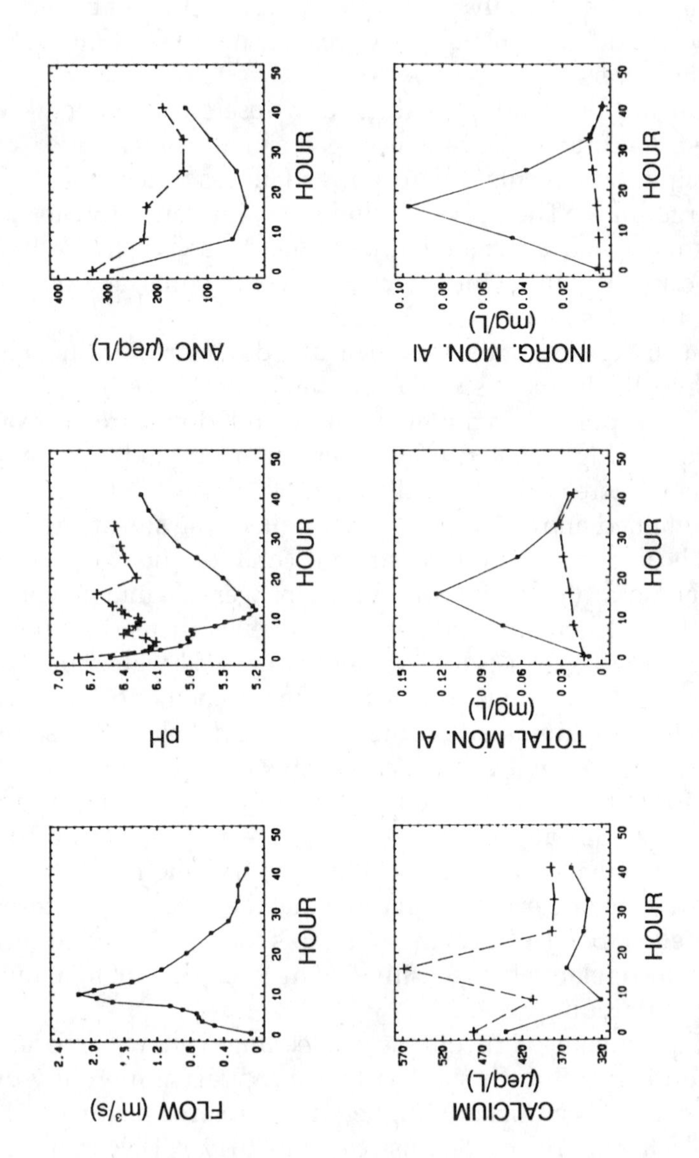

Figure C.3 Flow and water chemistry measured at upstream (•) and downstream (+) stations on Bacon Ridge Branch during a storm event in June 1987. (Source: Janicki and Greening, 1988a.)

cated that all anadromous fish populations were at low levels in Bacon Ridge Branch. In comparison, results from Mattawoman Creek showed relatively large populations of yellow perch, white perch, alewives, and blueback herring. Apparently, the fishery in Mattawoman Creek was less affected by adverse water quality.

In December 1988, a propane-powered doser was installed on Stocketts Run. Results of water quality and biological studies after liming are not yet available.

LIVING LAKES, INC.

Living Lakes, Inc., a not-for-profit aquatic fish restoration demonstration program, was started in 1986 with funding from private sources to demonstrate the technical and cost effectiveness of liming (Brocksen and Emler 1988). To date, an evaluation of seven liming technologies has been conducted on 37 lakes and 11 streams in eight states: Kentucky, Maryland, Massachusetts, Michigan, New York, Pennsylvania, Rhode Island, and West Virginia. The program includes ongoing water quality analysis for the lakes and streams treated and, if necessary, retreatment. Some projects were conducted in cooperation with other organizations, and results are reported elsewhere in this section, e.g., Electric Power Research Institute and Maryland studies.

Marcus (1988) evaluated results of pre- and postliming water quality for 22 lakes treated during 1986 and 1987 under the Living Lakes program (Table C.5). Levels of ANC, calcium, conductivity, dissolved inorganic carbon, and pH were significantly higher after liming. Only zinc concentrations were significantly lower in the posttreatment samples; this difference was attributed to zinc contamination of the preliming sample filters.

A general review of the data indicated that in addition to increases in ANC, calcium, conductivity, dissolved inorganic carbon, and pH, there was a tendency toward increases in nitrate and dissolved organic carbon; in contrast, aluminum, sulfate, and other metals were generally lower in posttreatment samples.

On the basis of the results of the discriminant analyses, Marcus (1988) suggested that of the five variables most closely related to liming, ANC was most useful for distinguishing between pre- and post-treatment. Its usefulness was partly explained by the significant correlation between ANC and several other variables—

Table C.5 Mean, Range, and Numbers of Samples (N) for Water Chemistry Variables Before and After Liming of 22 Living Lakes Study Lakes (Source: Marcus, 1988)

	Before Liming				After Liming			
Variable[a]	Mean	Min.	Max.	N	Mean	Min.	Max.	N
Aluminum, µg/L	40	BD[b]	200	112	40	BD	180	234
ANC, µeq/L[c]	25.4	-35.4	90.4	122	193.3	1.2	638.9	231
Calcium[c]	1.78	0.43	3.5	115	5.09	1.49	13.33	236
Cadmium	BD	BD	BD	112	BD	BD	BD	117
Chloride	5.98	0.20	20.76	15	21.02	0.21	70.90	36
Conductance, µS/cm[c]	39.49	BD	97.40	130	64.05	13.80	278.0	222
Dissolved inorganic carbon[c]	1.62	BD	9.11	137	2.91	BD	13.3	216
Dissolved organic carbon	2.96	0.91	11.86	128	2.91	BD	13.43	224
Iron	0.03	BD	0.83	113	0.05	BD	1.23	112
Potassium	0.51	0.24	1.05	16	0.45	BD	0.96	26
Magnesium	0.95	0.09	2.40	28	1.10	0.14	2.22	32
Manganese	0.06	BD	2.70	113	0.03	BD	1.45	105
Sodium	3.38	0.13	9.63	15	12.33	0.08	48.80	34
Ammonium	0.01	BD	0.11	15	0.04	BD	0.32	27
Nitrate	0.33	BD	1.71	112	0.35	BD	1.90	117
Lead	0.01	BD	0.03	113	0.01	BD	0.03	117
pH[c]	5.51	4.44	6.99	128	6.62	5.40	8.63	226
Silica	1.60	BD	6.65	14	1.82	BD	6.49	24
Sulfate	5.67	BD	8.29	112	5.58	0.62	7.61	117
Total phosphorus	BD	BD	0.05	112	BD	BD	0.06	217
Zinc[c]	0.03	BD	0.21	113	BD	BD	0.09	118

[a] All units in mg/L except where indicated.
[b] BD = below detection limit.
[c] Significant difference ($p \leq 0.05$) between pre- and posttreatment means based on Mann-Whitney U test indicated by underlined values.

calcium, conductivity, dissolved inorganic carbon, dissolved organic carbon, pH, total nitrogen, and zinc. The overall comparison of pre- and postliming water quality suggested that the liming program had been successful in improving water quality for fish.

Appendix D

Results of Selected Mitigation Projects Outside the United States

The physical, chemical, and biological effects of selected liming experiments conducted by mitigation programs in Canada, Sweden, Norway, and Scotland have been reported. Unlike the more comprehensive descriptions of all major mitigation programs in the United States (Appendix C), the projects described here only highlight one or two important projects in each country.

CANADA

In 1981, the Ontario Ministers of the Environment and Natural Resources began a 5-year investigation of the feasibility of neutralization as a technique to rehabilitate acidified lakes and to protect acid sensitive lakes (Dodge et al., 1988). An acidic lake, Bowland Lake, and an acid-sensitive lake, Trout Lake, were limed by fixed-wing aircraft in 1983 and 1984, respectively. Another acid sensitive lake was tested for water chemistry and fish but was not treated, and a fourth lake was selected as a nonacidic control lake for bioassay experiments. A summary of the results of experiments conducted from 1981 to 1987 was issued by Dodge et al. (1988). Specific aspects of the work were recently reported in the *Canadian Journal of Fisheries and Aquatic Sciences* (Gunn et al., 1990; Jackson et al., 1990; Keller et al., 1990a,b; Molot et al., 1990a,b).

Trout Lake was nearly three times larger than Bowland Lake, but had positive alkalinity values and much lower aluminum concentrations (Table D.1).

After treatment with limestone, Bowland Lake underwent

Table D.1 Morphometric and Chemical Characteristics of Bowland Lake and Trout Lake in Ontario, Canada, Before Limestone Treatment (Source: Dodge et al., 1988)

Variable	Bowland Lake[a]	Trout Lake[b]
Surface area, ha	108.0	290.0
Maximum depth, m	28.0	37.8
Mean depth, m	7.0	11.0
Retention time, years	2.0	2.7
pH[c]	4.78 to 5.76	5.64 to 6.20
ANC, μeq/L[c]	-17.8 to 9.4	15.6 to 24.6
Total aluminum, μg/L[c]	100 to 160	16 to 63
Conductance, μS/cm[c]	35 to 42	28 to 29
True color, Hazen units[c]	1.5 to 7.4	7.0

[a] Values measured before liming from June 1982 to October 1983.
[b] Values measured before liming in August 1983.
[c] Samples collected from the epilimnion at midlake.

major changes in water chemistry; pH increased in pH from 4.9 to 6.7 and alkalinity from -6 to 90 μeq/L. Total aluminum concentrations decreased gradually from 130 to 30 μg/L. Transparency decreased in the lake; Secchi depth decreased from 6.6 m to 5.4 m a year later.

Trout Lake, being initially higher in pH and ANC, changed much more subtly; pH increased from 5.8 to 6.6 and ANC from 20 to 80 μeq/L. Aluminum concentrations did not change appreciably after neutralization; concentrations ranged from 30 to 35 μg/L before and after liming.

Reacidification of both lakes was monitored. Bowland Lake lost 40% of its ANC from August 1983 (when it was treated) to March 1986. Trout Lake lost about 25% of its alkalinity during an 18-month period after treatment. Calcium and pH changes in the lakes during these periods (Figure D.1) were compared to the Sverdrup reacidification model (Dodge et al., 1988). Spring snowmelt pH depressions for both lakes were observed to be more severe than those predicted by the model, which was developed for Swedish climatic conditions.

Biological responses evaluated in Bowland Lake included benthic algae, macrophytes, phytoplankton, zooplankton, benthic invertebrates, and fish. Only fish community responses were reported for Trout Lake.

One year after neutralization, the acidophilic algae in Bowland

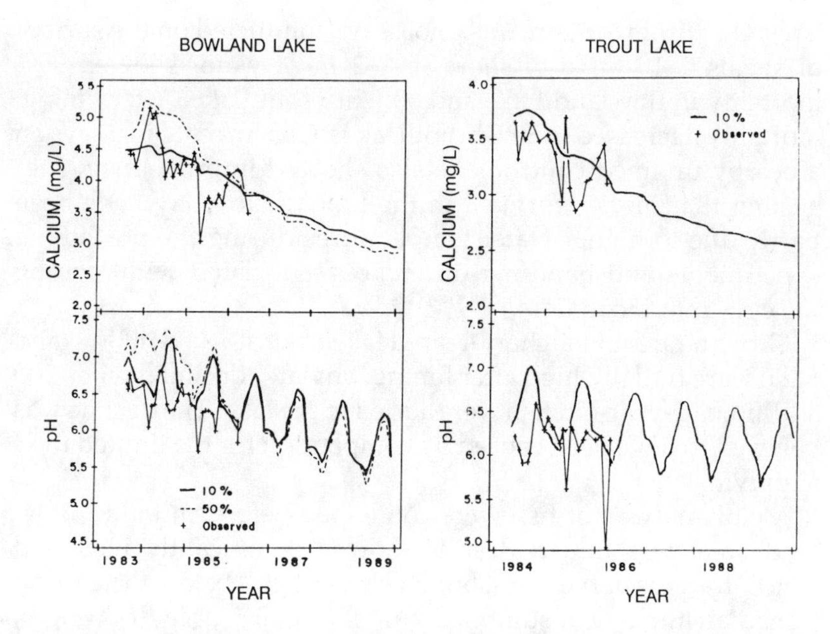

Figure D.1 Calcium and pH changes in Bowland Lake (left) and Trout Lake (right) after limestone treatment (+) and comparison with results of the Sverdrup reacidification model assuming that the settled limestone covered 10% (solid line) or 50% (broken line) of the lake bottom. (Source: Booth et al., 1987.)

Lake before treatment decreased dramatically. Algal abundance returned to preliming values 3 years after liming, but species composition were more typical of an acid-sensitive rather than an acidic lake. The aquatic macrophyte community was unaffected by the liming treatment. Phytoplankton before liming was dominated by *Rhabdoderma*, a blue-green alga. Species composition after liming became more diverse, becoming dominated by species more typical of nonacidic systems. Similar Rotifer communities changed similarly: dominance changed from the more acid-tolerant *Keratella taurocephala* to *K. cochlearis* and *Polyarthra* sp.

Crustacean zooplankton biomass decreased in Bowland Lake after liming, suggesting predation by fish and phantom midges. Similarly, decreases in benthic macroinvertebrate biomass and organism size after liming were believed to be attributable to predation by fish.

Bioassays with lake trout eggs were conducted in both lakes before and after liming to determine changes in survival rates as a result of liming treatments. Bioassays conducted on the surface of shoals 2–10 m from shore in 1–2 m of water showed 95% mortality in Bowland Lake and 60% in Trout Lake. After liming, mortality decreased to 30% in both lakes. Concurrent bioassays on a nearby unlimed control lake also showed improvement, suggesting that higher survival in the limed lakes may have been partly due to higher water temperatures during the postliming experiments and handling differences associated with loading the eggs into incubators.

Growth rates of hatchery-reared lake trout stocked in Bowland Lake were initially high after liming, but later declined when 200 additional 2-year-old fish were added to the lake. The decline was believed to have been caused by competition for the limited food supply.

Metal analyses of fish were conducted before and after Bowland Lake was limed. Results showed significantly increased concentrations of mercury, but levels were still below the current Canadian human consumption guideline of 0.5 μg/g dry weight, and levels decreased to preliming concentrations 1 year after neutralization. It was postulated that changes were due to natural annual variations in mercury concentrations.

SWEDEN

Sweden has the largest and most comprehensive aquatic liming program in the world. Both operational and experimental liming programs have been conducted with funding from the Swedish government (Dickson, 1985; Nyberg and Thornelof, 1988).

As an example of Swedish liming results, the effects of liming Lake Gardsjon in April 1982 are described here. This was the most comprehensive individual study of lake liming yet conducted in Sweden. A detailed evaluation of the physical, chemical, and biological effects of the addition of powdered limestone to the lake surface was conducted by Swedish researchers in the Lake Gardsjon watershed over an 8-year period. Before this, the lake was fishless.

Unlike most other liming studies where lakes were stocked soon after treatment, fish were introduced into the lake after 4 years of intensive physical, chemical, and biological evaluations after liming. This is an important consideration because direct biological effects of liming are often masked by fish predation. Knowledge of what happens to aquatic biota in both situations is useful.

Lake Gardsjon (Figure D.2) is in southwest Sweden, 50 km north of Gothenburg. The watershed area includes three other major lakes, one of which (Lake Stora Hastevatten) was monitored as a control (Hultberg and Nystrom, 1988).

Preliming data collection began in 1979 and the lake was treated in 1982 by the addition of slurried limestone by boat (Dickson, 1988). The lake was treated again in late 1985, but results of this treatment were not reported. Lake Gardsjon is larger than the control lake, Lake Stora Hastevatten (Table D.2); the control lake had similar water chemistry before treatment (Table D.3).

The liming treatment resulted in an initial increase in pH to 7.9 (Figure D.3); the next summer it was about 7.2. The pH decreased during two high discharge periods, to pH 6.5 in January 1984 and to pH 5.3 in April 1985 (Figure D.3). Both of the decreases were temporary; pH returned to about 7 after several weeks.

The transparency decreased dramatically for 4–5 weeks after the liming due to suspended limestone, and thereafter remained lower than before liming. This was considered to be partly because of increased color and mostly due to an increase in phytoplankton biomass. The increase in phytoplankton was represented by the higher chlorophyll *a* concentrations observed after liming (Table D.3).

Concentrations of dissolved organic carbon increased sharply after liming. Broberg (1988) attributed the increases to three factors: mineralization of the acidic benthic algae after liming, followed by increased dissolved organic carbon; increased algal respiration at the sediment surface with subsequent release of organic carbon; and reduced photochemical degradation of organic matter after liming.

Overall, total phosphorus and nitrogen remained relatively unchanged after liming (Broberg, 1988). However, nitrogen and phosphorus decreased in the control lake. The reason cited for the lack of corresponding decreases in the limed lake was the release of nutrients from decomposition of phytoplankton.

Total aluminum concentrations decreased significantly after liming, from about 300 to 200 μg/L.

Gahnstrom (1988) evaluated the decomposition of organic

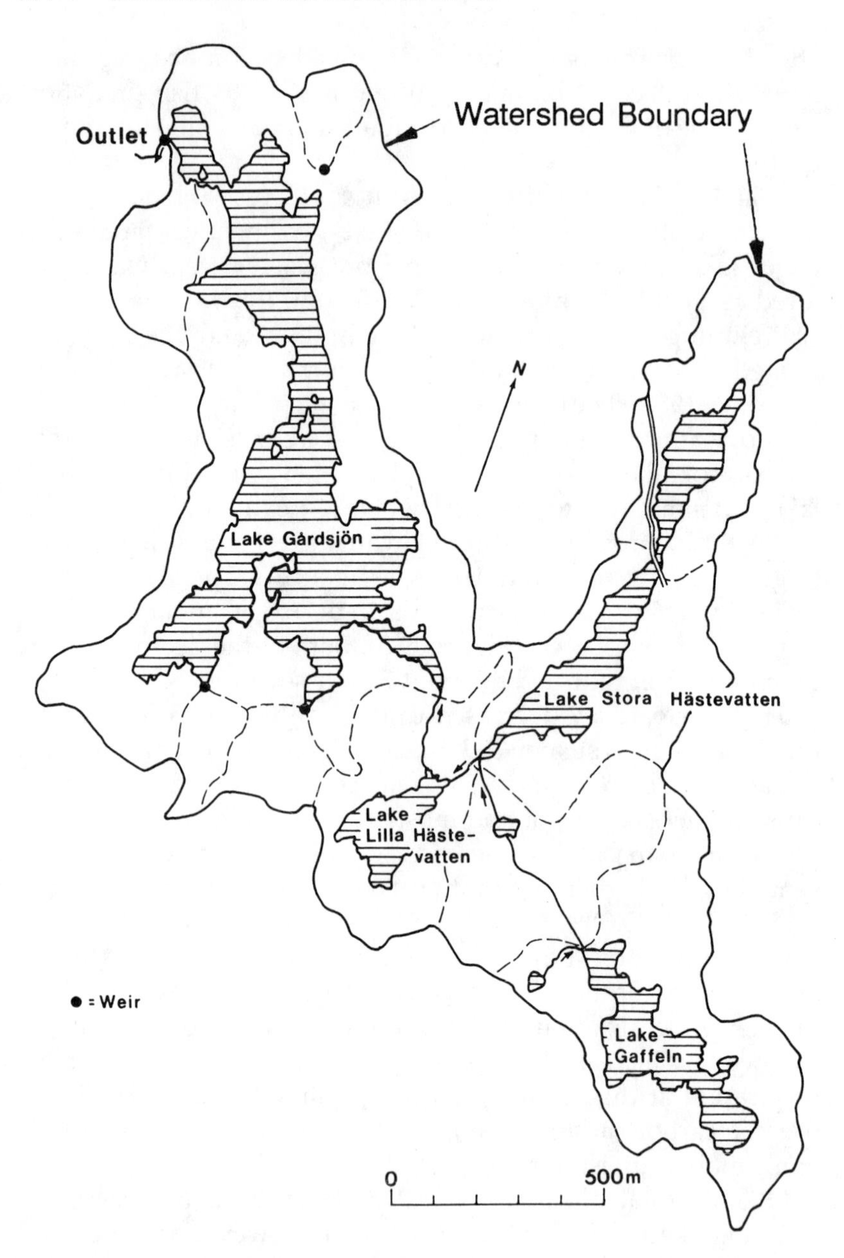

Figure D.2 The Lake Gardsjon watershed showing the location of weirs at selected subwatersheds. (Source: Johannson and Nilsson, 1988.)

Table D.2 Morphometric Characteristics of Lake Gardsjon and Lake Stora Hastevatten (Source: Hultberg and Nystrom, 1988; Gahnstrom, 1988)

Variable	Lake Gardsjon (Treated)	Lake Stora Hastevatten (Control)
Surface area, ha	31.2	8.4
Volume, m^3	1,500,000	311,000
Maximum depth, m	18.5	9.3
Mean depth, m	4.9	3.8
Retention time, years	1.36	NR[a]

[a] Not reported.

material and microbial activity in the lake sediment by measuring oxygen uptake. A decrease in sediment oxygen uptake observed in Lake Gardsjon after liming lasted only 2 weeks and was not seen in the control lake. The decrease was believed to be directly due to an increase of 2 pH units in the interstitial water, resulting in stress to benthic organisms. Deposition of calcium, aluminum, and other metals was suggested as an additional stress.

After this initial decrease in sediment oxygen uptake, there was an increase that lasted several months and was explained by the decomposition of organic material from decaying biomass.

Before liming, the phytoplankton community structure was dominated by large dinoflaggelate species such as *Peridinium inconspicuum* and *Gymnodinium* spp. After an initial decline in number of species and biomass, both steadily increased in diversity and biomass. The trend continued until 1985, when a massive bloom of *Cosmocladium perissum* occurred.

The dominant macrophyte before liming was *Sphagnum subsecundum.* This species almost vanished after liming. *Potamogeton alpinus,* a macrophyte absent before liming, was seen 1 year after treatment. The disappearance of sphagnum was the most significant change in the distribution, biomass, or production of macrophytes after liming. Figure D.4 illustrates the changes in distribution of macrophytes in the lake due to liming. Total biomass was reduced about 50% and net production about 70% after treatment.

Table D.3 Mean Physical and Chemical Characteristics of Lake Gardsjon Before and After Liming and Comparison with a Control Lake (Source: Gahnstrom, 1988)

	Lake Gardsjon				Lake Stora Hastevatten			
	Before[a]		After[b]		Before[a]		After[b]	
Variable	Mean	Range	Mean	Range	Mean	Range	Mean	Range
pH	4.7	4.5–4.8	NR[c]	5.5–7.9	NR	5.0–5.5	NR	4.9–5.6
Transparency, m	8.9	6.5-13.0	NR	1.4-7.9	NR	NR	NR	NR
Color, mg Pt/L	3.8	0.5-11.6	5.6	0-14.1	4.4	0.9-14.8	4.0	0-9.5
Alkalinity, μeq/L	0	0	NR	0-540	0	0	0	0
Calcium, mg/L	1.8	1.6-2.4	NR	2.7-13.5	2.0	NR	2.0	NR
Aluminum, μg/L	300	100-500	200	NR	300	NR	NR	NR
Total phosphorus, μg/L	4.8	1.6-9.8	5.1	3.0-8.9	3.8	1.1-7.1	3.3	1.3-9.2
Total nitrogen, μg/L	373	232-552	369	226-489	369	233-512	255	111-373
Dissolved organic carbon, mg/L	2.3	0.5-3.8	4.1	1.3-8.7	3.7	2.4-5.3	3.4	0.7-4.7
Chlorophyll *a*, μg/L	0.64	0.07-1.63	3.40	0.13-14.65	1.00	0.19-2.40	0.90	0.10-2.90

[a] Before limestone was added to Lake Gardsjon.
[b] After limestone was added to Lake Gardsjon.
[c] Not reported.

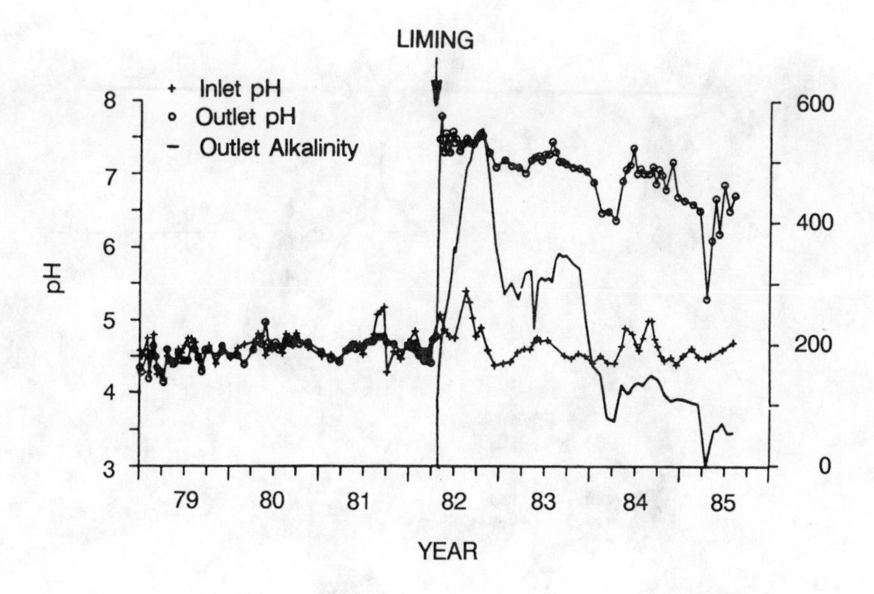

Figure D.3 Inlet and outlet pH and alkalinity in Lake Gardsjon before and after liming treatments. (Source: Hultberg and Nystrom, 1988.)

NORWAY

Although Norway faces many of the same surface water acidification problems faced by Sweden, its government-supported operational program is somewhat smaller (Rosseland and Hindar, 1988). Norway also has a comprehensive research and development program. The liming of Lake Hovvatn, which began in 1981, has been the most comprehensive of the Norwegian research liming projects; results of the liming were reported by Wright (1984, 1985), Rosseland and Hindar (1988), and Langvatn (1989).

Liming of a watershed in 1983 for mitigation of acidity in Lake Tjonnstrond was also conducted under the Norwegian Liming Project. The results of this program are also included here.

Lake Hovvatn

Lake Hovvatn and a smaller upstream pond, Pollen, were limed in 1981 with 200 and 40 tonnes, respectively, of powdered limestone distributed on the ice and along the shore of the lake

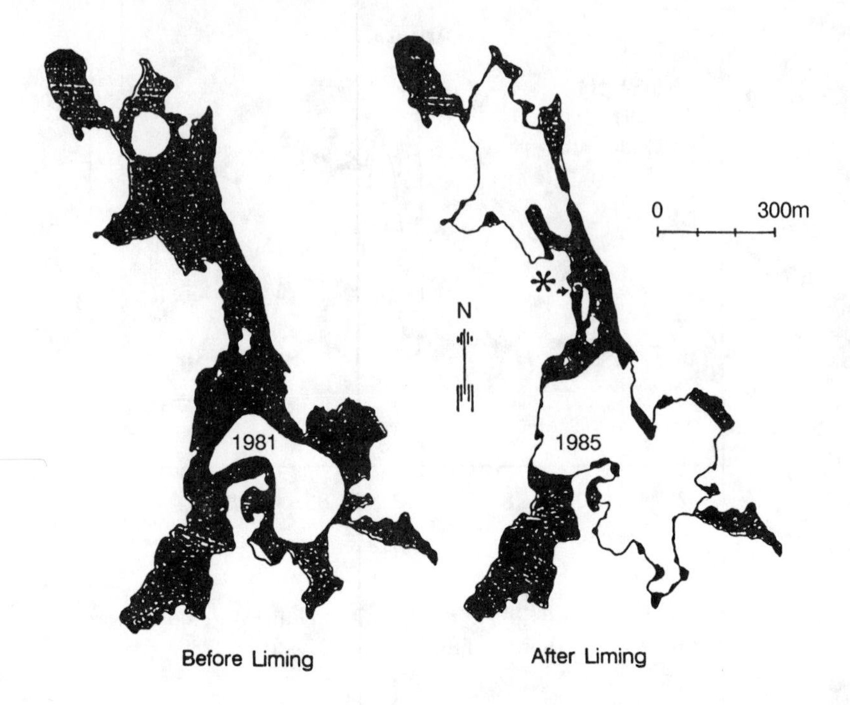

Figure D.4 Distribution of macrophytes (shaded areas) in Lake Gardsjon before liming (1981) and 3 years after liming (1985). All species are included. The asterisk indicates the site of colonization of the new species *Potamogeton alpinus* in 1983. (Source: Grahn and Sangfors, 1988.)

(Figure D.5). Another lake in the watershed, Lake Lille Hovvatn, was not limed and served as a control.

The research involved evaluation of the water chemistry and benthic and fishery biology in the lake until reacidification. Lake Hovvatn was relimed in 1987 with 91 tonnes of limestone after reacidification had proceeded to the point where the trout population had decreased significantly.

Lake Hovvatn is much larger and deeper than Pond Pollen (Table D.4). Before liming, both lakes were characterized as highly acidic (pH 4.5), with negative alkalinity, low calcium, and high aluminum (Table D.5).

Limestone applications during winter 1981 resulted in an increase in pH to about 6.7 in Hovvatn and 7.5 in Pollen after ice break-up. Because lime application included shore liming, the initial dissolution (2 weeks after ice break-up) was only 14%. After 1 year, the pH was 5.66 in Hovvatn and 6.35 in Pollen.

Among trends in pH and calcium for the two lakes and the

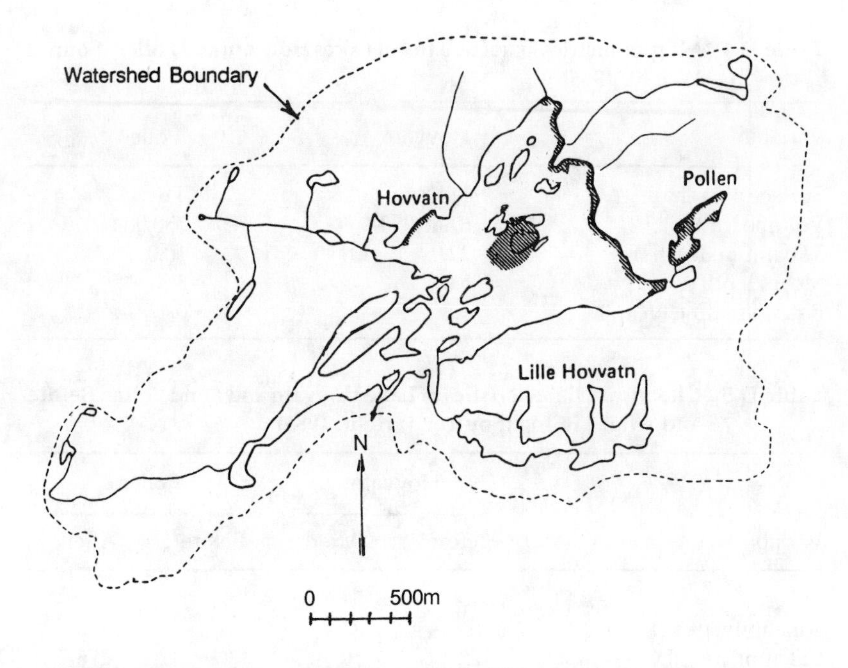

Figure D.5 The Lake Hovvatn watershed, showing the upland pond, Pollen, and the control lake, Lille Hovvatn. Hatched areas show where limestone was applied. (Modified from Rosseland and Hindar, 1988.)

reference lake (Figure D.6), depressions in pH of more than one unit in the limed lakes corresponded to periods of snowmelt and the flow of acid water beneath the ice cover; the pH and calcium in the control lake were almost unchanged throughout the period. Some decreases in pH were observed for the control lake during snowmelt periods, but the initially low pH of about 4.5 did not vary more than about 0.2 unit.

The aluminum concentrations were reduced in the limed lakes to about half the preliming values 1 year after liming. These concentrations gradually increased as reacidification progressed. Total organic carbon and total phosphorus doubled after 1 year. These carbon and phosphorus increases were not believed to be attributable to phytoplankton growth; rather a combination of factors was suggested. The phosphorus increases were believed to be due to trace levels of phosphorus in the limestone and possibly due to the release of phosphorus from sediments. The variations around the phosphorus values, however, are probably greater than the actual difference between the before and after values.

Table D.4 Morphometric Characteristics of Lakes Hovvatn and Pollen (Source: Wright, 1985)

Variable	Hovvatn	Pollen
Surface area (ha)	114	4.6
Volume (m^3)	6,400,000	150,000
Maximum depth (m)	22.0	10.0
Mean depth (m)	5.6	3.3
Retention time (years)	0.8	0.4

Table D.5 Chemical Characteristics of Lake Hovvatn and Pond Pollen Before and After Liming (Source: Wright, 1985)

	Hovvatn		Pollen	
Variable	Before[a]	After[b]	Before[a]	After[c]
pH	4.45	5.66	4.47	6.35
Alkalinity, μeq/L	0	28	0	180
Calcium, μeq/L	22	107	22	276
Total aluminum, μeq/L Al^{3+}	34	12	34	22
Total phosphorus, μg/L	3	7	5	9
Total nitrogen, μg/L	400	550	300	570
Total organic carbon, mg/L	2.2	3.7	3.2	5.3

[a] Samples collected on May 21, 1980, at 1-m depth.
[b] Samples collected 1 year after liming on April 21, 1982, at 5-m depth.
[c] Samples collected 1 year after liming on April 21, 1982, at 6-m depth.

The composition of invertebrates in the lakes before treatment was typical for acidic lakes; oligochaetes were relatively abundant in the littoral zone, but not at depths of 5–10 m. Diversity and density of invertebrate species were low before liming. The response of invertebrates to liming was slower in Hovvatn than in Pollen. Oligochaetes increased in number at all depths during the first year after liming in Pollen, but not until the second year in Hovvatn. The difference in response was ascribed to differences in bottom substrate and possibly limestone particles on the shore of Hovvatn.

An increase in chironomids in the benthos of both lakes was delayed. This delay was attributed to the accumulation of available food sources. A dramatic decrease in 1985 was attributed to depletion of food sources and possibly reduced availability of organic matter as the lake became reacidified.

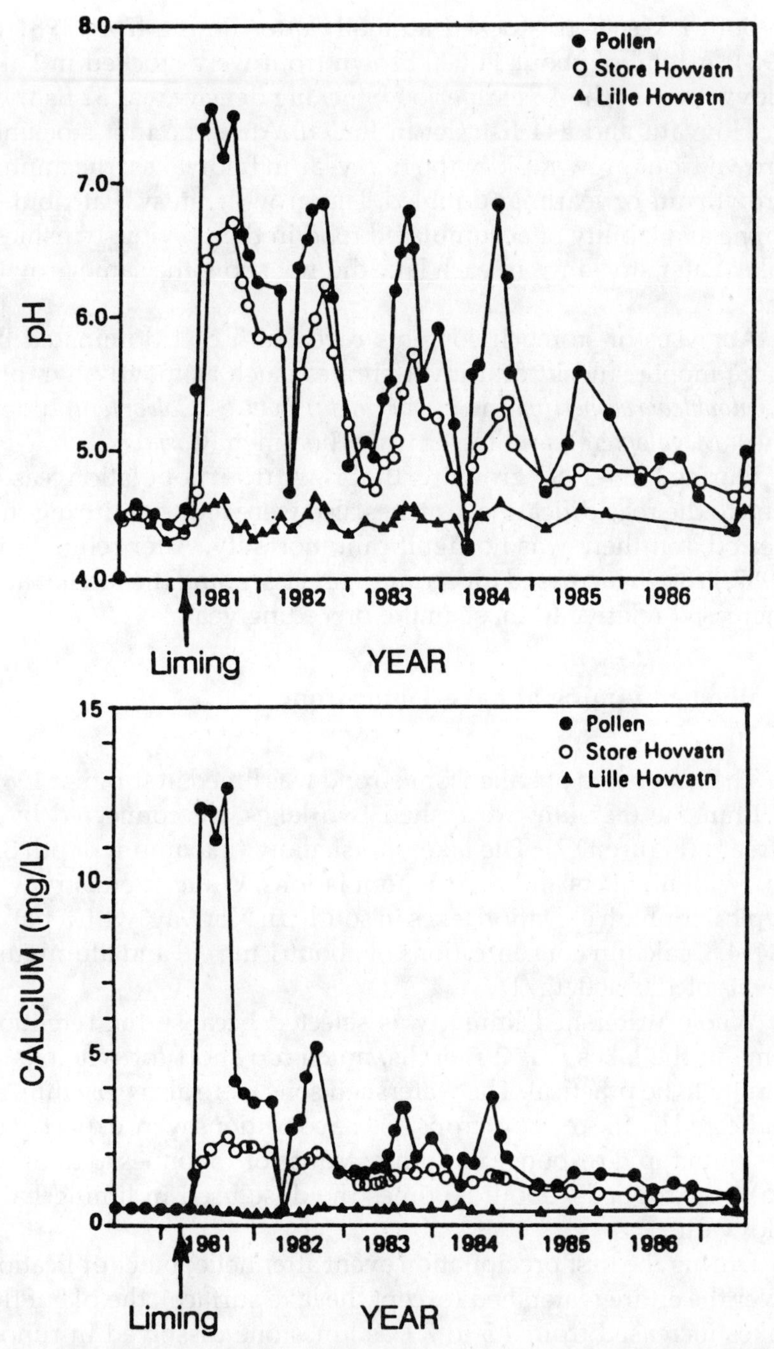

Figure D.6 Trends in pH and calcium of the treated lakes (Hovvatn and Pollen) and the control lake (Lille Hovvatn) from just before liming in 1981 through 1987. (Modified from Rosseland and Hindar, 1988.)

Both lakes were stocked annually after liming from 1981 to 1984. A total of about 11,000 brown trout were stocked in Lake Hovvatn over the 4-year period. Stocking density was 32 fish/ha in Hovvatn and 244 fish/ha in Pollen. After the first stocking, brown trout grew rapidly the first year in both lakes; the annual growth rate of yearlings doubled. This growth rate was attributed to the availability of accumulated food in the previously fishless lake. Later stockings in each lake did not show the same growth rate.

Analyses of stomach contents revealed a diet dominated by large mobile and littoral invertebrates, such as mayfly nymphs (*Leptophlebia vespertina* and *L. marginata*), beetles (*Coleoptera*), alder-flies (*Megaloptera*), and lesser water boatmen (*Corixidae*).

During 1986–1987, growth of the brown trout population ceased due to the reacidification. Catches also were reduced during this period, but there was no significant mortality. After reliming in 1987, growth increased for certain age classes, and the catches also increased relative to those in the preceding year.

Watershed Liming at Lake Tjonnstrond

The watershed of Lake Tjonnstrond was limed in summer 1983. Within the 0.25-km^2 watershed two lakes are connected by a stream (Figure D.7). The lakes are shallow (maximum depth 3.5 m), with no inlets and no fish populations. Water chemistry was typical for high elevation lakes in southern Norway, with a pH of 4.4–4.8, calcium concentrations of about 1 mg/L, and aluminum levels of 100–400 μg/L.

Whole watershed liming was selected because the retention time in the lakes was 2 months, much too short for whole-lake liming to be practical. The watershed soils were also very thin, as indicated by the rock outcrops. These conditions were expected to result in rapid response in runoff. A dose of 75 tonnes (3 tonnes/ha) was applied—about 50 times the dose used in liming Lake Hovvatn.

During the first precipitation event after helicopter application over the entire watershed (except the lake surface), the pH of the lakes increased from 4.5 to 7.1 as limestone dissolved in runoff water (Figure D.8). Decreases in pH below the high of pH 7.1 were observed during later storm events, particularly in spring, although the pH remained above that believed critical for the fish

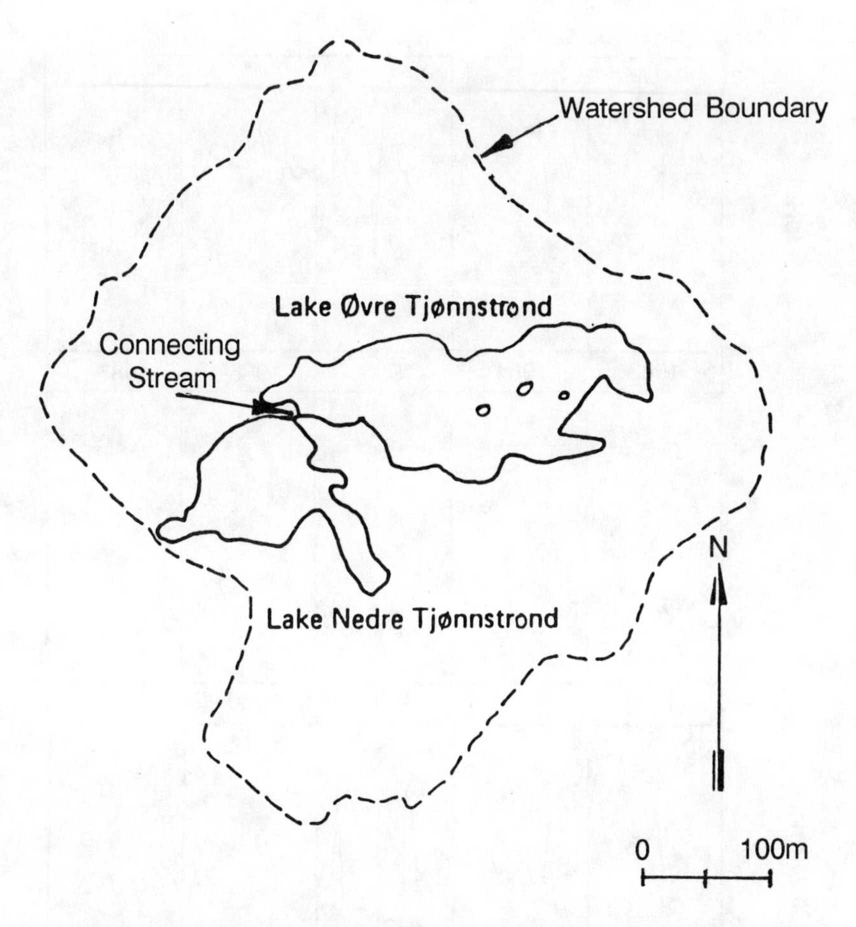

Figure D.7 The Lake Tjonnstrond watershed. (Modified from Rosseland and Hindar, 1988.)

population introduced into the lakes 4 months after liming (generally above pH 5.8).

Figure D.8 also shows that aluminum and calcium levels over the 5-year period after watershed treatment compared to those in a nearby lake that served as a control. Aluminum concentrations were much lower than those in the control lake (Lake Storgama) and were lower than those typically seen in whole-lake liming experiments in Norway. This was thought to be due to the continued input of high aluminum water after whole-lake liming. Treatment of the watershed, on the other hand, is effective in preventing high aluminum in runoff entering the lake. Calcium increased from preliming levels below 1 mg/L to about 4–5 mg/L, then decreased steadily to about 2 mg/L by 1988.

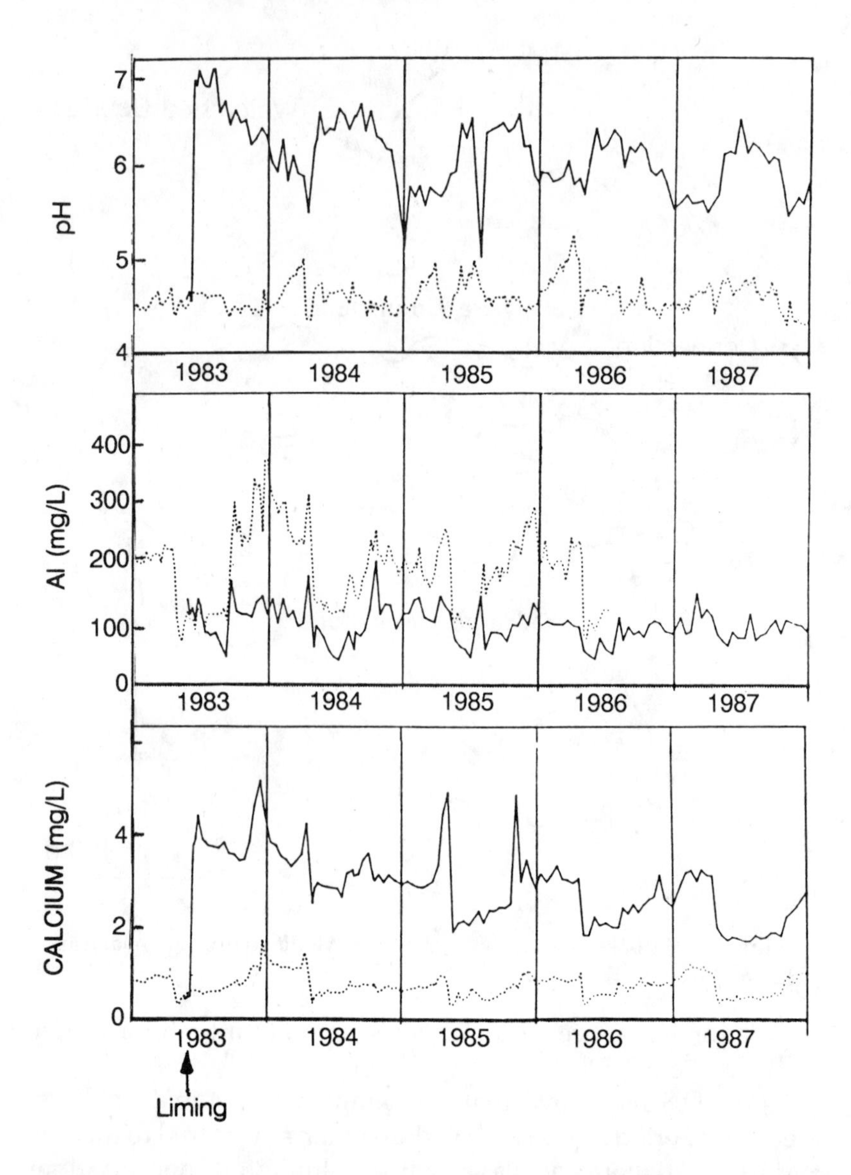

Figure D.8 The pH, calcium, and aluminum at the outlet of Lake Tjonnstrond (solid line) compared to the control lake Storgama (broken line). (Modified from Rosseland and Hindar, 1988.)

The limestone dose selected for the watershed was calculated to maintain acceptable water quality for 30 years. Although the researchers stated that they did not expect the effects of liming to last as long as calculated, they were optimistic about relatively long-lasting water quality improvement. Four months after liming, 125 brook trout were stocked in Lake Nedre Tjonnstrond and 125 brown trout in Lake Ovre Tjonnestrond. Successful survival and growth were seen, with very little migratory activity. As is typical for the species, brook trout exceeded the brown trout in specific growth.

SCOTLAND

The focus of liming research in Scotland has been experimental, with emphasis on two lakes in Scotland, Loch Dee (Burns et al., 1984; Tervet and Harriman, 1988) and Loch Fleet (Brown, 1988; Dalziel et al., 1988). The liming of Loch Fleet in 1986 and 1987, which has been the most comprehensive program in Scotland, is described here.

Loch Fleet is a 17-ha, high-elevation (340 m), fishless lake in southwest Scotland. In April 1986, a total of 350 tonnes of limestone powder and slurry were manually applied to three subwatersheds: sector IV was treated with a slurry at the rate of 20 tonnes/ha; sector VI was treated at 30 tonnes/ha with a powder; and sector VII was treated with a powder at 10 tonnes/ha in wetland areas covering about 25% of the subwatershed (Figure D.9).

Before treatment the lake pH ranged from 4.1 to 5.0, calcium concentrations were about 1 mg/L, and aluminum ranged from 120 to 290 µg/L. The water chemistry in all three subwatersheds reacted similarly to the different types of treatments. The pH of the water increased rapidly and maintained values near 7 during the period reported from April 1986 to December 1987. Calcium concentrations increased dramatically from about 1 to 30 mg/L after liming and total aluminum decreased from 150 to 200 µg/L in each of the subwatersheds to concentrations that were frequently less than 100 µg/L.

As expected, the water chemistry at the lake outlet (Figure D.10) responded more slowly to liming treatments than did the individual subwatersheds, as represented by sector VII (Figure D.11). Also, the changes were less dramatic because some subwatersheds were not treated and continued to contribute acidic water to the lake. The pH increased from 4.5 to about 6.5 after 2 months. Calcium concentrations increased from about 1 to 3 mg/

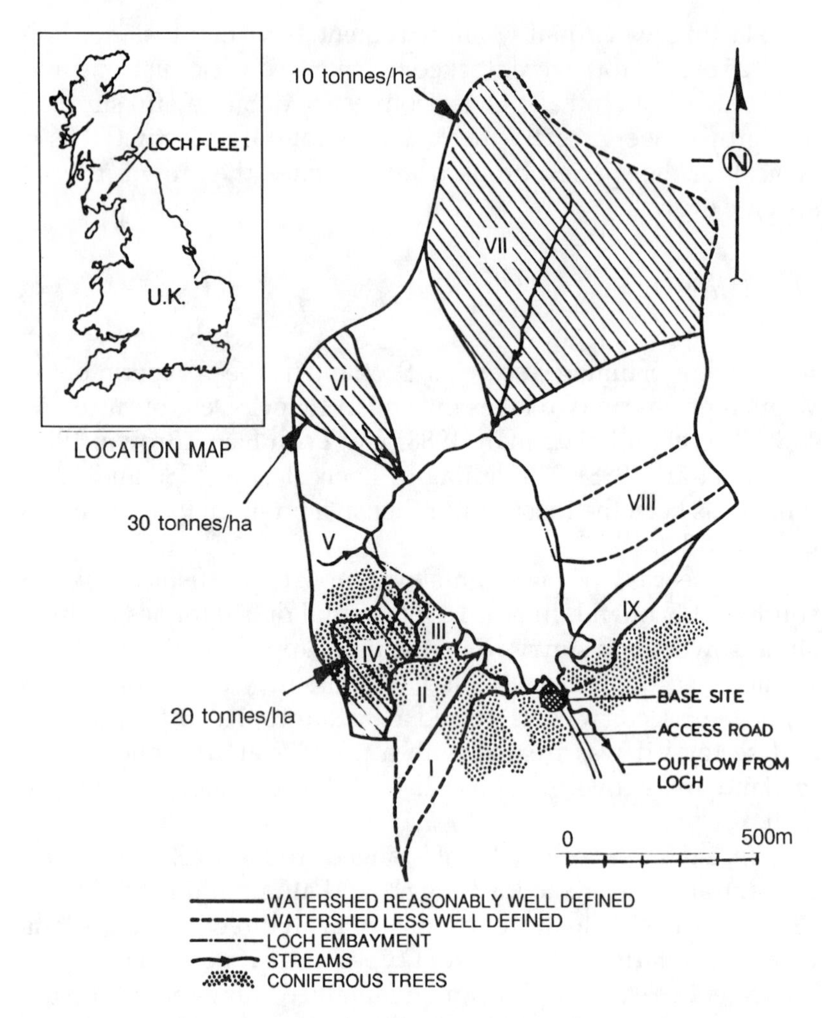

Figure D.9 Loch Fleet watershed showing the subwatersheds treated and limestone application rates. (Modified from Brown et al., 1988.)

L over the same period. The decreases in total aluminum were smaller than those in the subwatersheds, although labile (inorganic) monomeric aluminum decreased from about 50 to about 10 μg/L.

A different set of subwatersheds were treated in 1987 with much lower application rates (5 tonnes/ha) for a total application to the three subwatersheds of 90 tonnes. The much smaller limestone dosage was much less successful in improving the quality

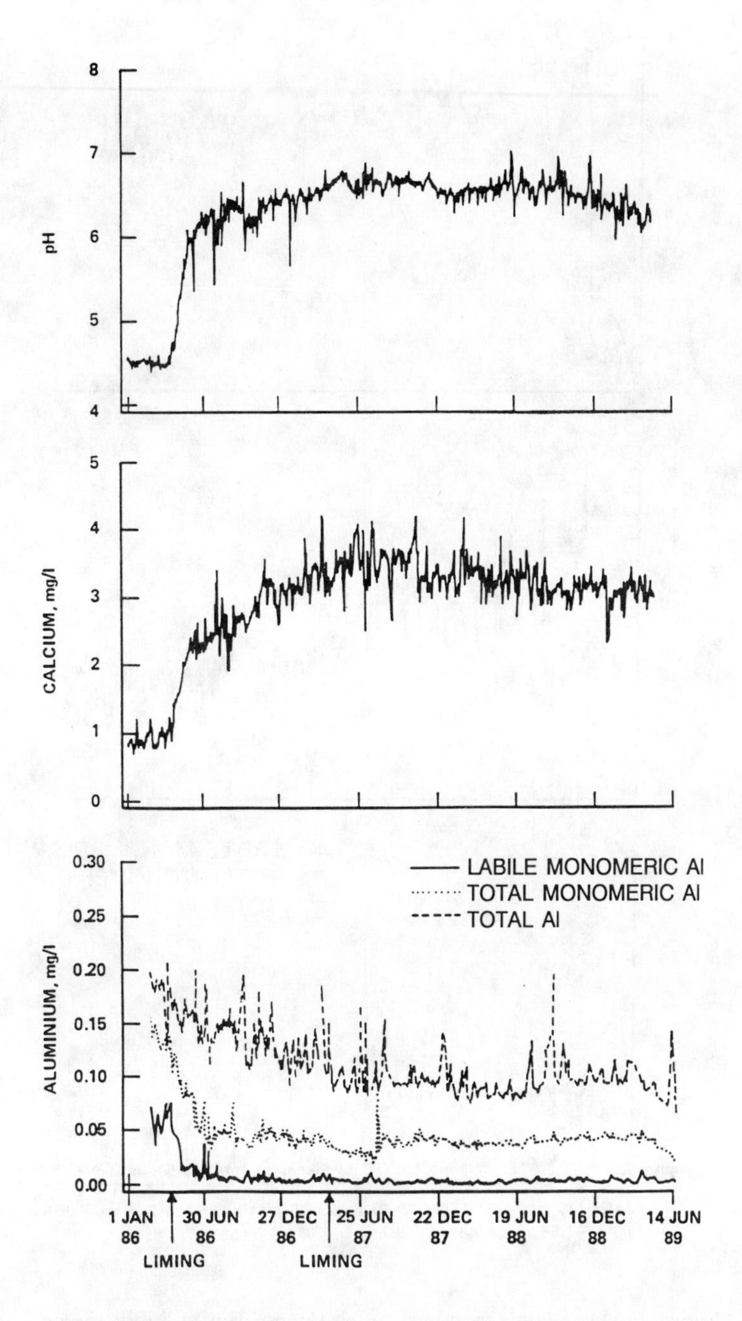

Figure D.10 Water chemistry at the outlet of Loch Fleet, showing when limestone was applied to subwatersheds in 1986 and 1987. (Modified from Brown et al., 1988.)

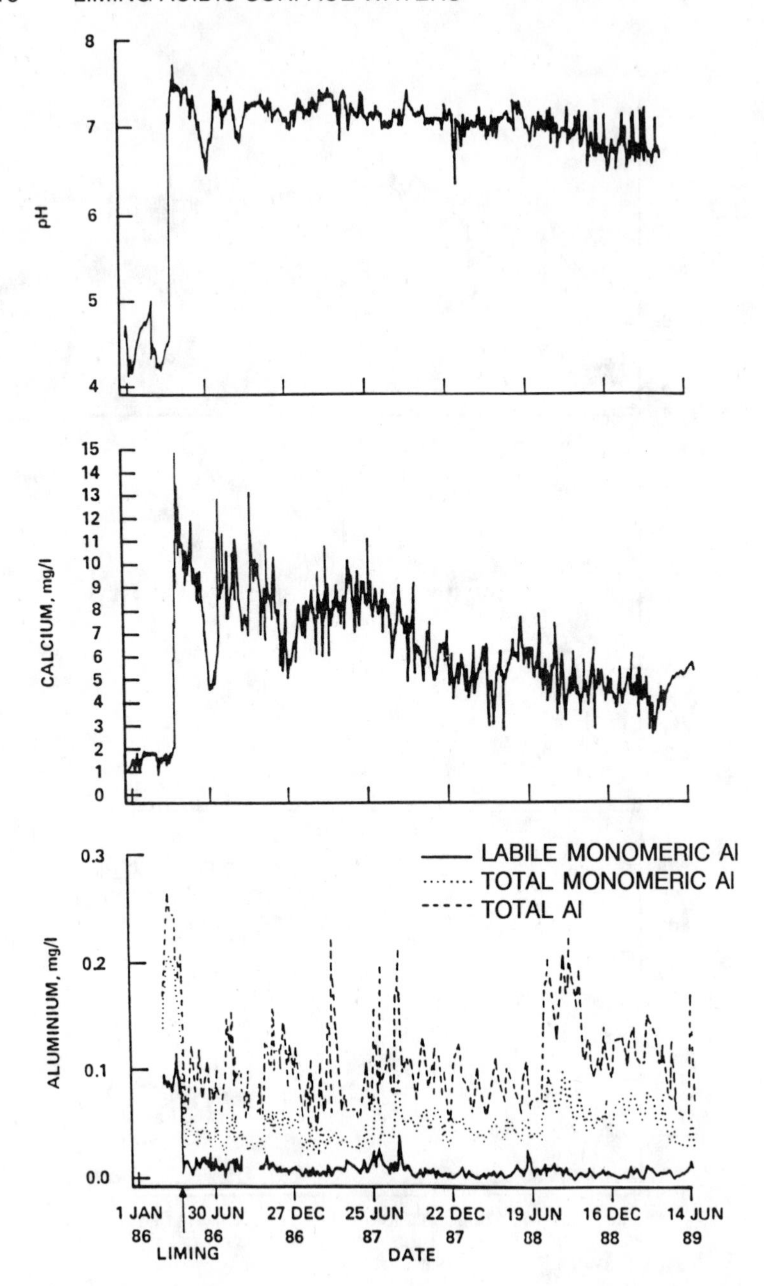

Figure D.11 Water chemistry of drainage from a subwatershed (Sector VII) of Loch Fleet, showing when limestone was applied. (Modified from Brown et al., 1988.)

of water leaving the subwatersheds. The pH was already at 7, however, and the calcium at 3 mg/L.

Effects to the terrestrial vegetation were studied intensively by comparing the results of surveys conducted before and after liming. In general, the dominant vegetation before liming, *Calluna*, was replaced by *Molinia*. Wetland areas were dominated by sphagnum before treatment. One year after limestone application, all *Sphagnum papillosum*, the dominant species, that had been exposed to limestone were dead. Certain other species were more tolerant of the limestone treatments, such as *S. inundatum*, *S. recurvum*, and *S. capillifolium*.

Aquatic macrophytes in the lake were also affected by the liming treatments. More than 90% of the sphagnum died back 18 months after the liming. A previously absent macrophyte, *Potamogeton polygonifolius*, became established in shallow water. Other species, such as *Isoetes echinospora*, *Juncus bulbosus*, and *Utricularia intermedia*, became more abundant after liming.

For aquatic invertebrates, liming resulted in general decreases in density during the season after treatment, probably due to the loss of acidophilic species. Samples collected near shore gradually returned to preliming levels, but numbers in deep water did not increase significantly. The abundance and diversity of invertebrates in the inlet and outlet streams increased after liming.

Brown trout were stocked in the lake after investigations revealed that water quality conditions after liming were suitable for a self-sustaining trout population. A total of 300 fish (about 18/ha) were introduced into the lake in May 1987. Trout fry were later seen in inlet and outlet streams, indicating that reproduction had occurred.

Index